FOOD CONSERVATION

Attila the Hun riding his beef tender, drawn by Ingrid Lowzow (see p 215).

FOOD CONSERVATION

edited by

Astri Riddervold
and
Andreas Ropeid

Prospect Books
1988

COVER ILLUSTRATION. This is a drawing of a girl shaking yoghurt in a goatskin to make butter (see p 189) by Tom Siwinski.

ISBN 0 907325 40 8

published by Prospect Books Ltd
45 Lamont Road
London SW10 0HU

distributed in the USA by
The University Press of Virginia
Box 3608 University Station
Charlottesville VA 22903

cover design by Ingrid Lowzow

printed and bound by
Smith Settle
Otley Mills
Otley, West Yorkshire

TABLE OF CONTENTS

TABLE OF CONTENTS

INTRODUCTION

In his speech of welcome to the Seventh International Ethnological Food Research Conference, the president of the conference, Dr. Anders Salomonsson of the University of Lund, emphasized that the important task in ethnological research is: 'To study man as a cultural being, and not only his cultural products.' He also observed that studies of food and eating habits constitute one of the best starting points for the study of man, since no other topics do more to reveal people's values, attitudes and views.

The subject of the conference was 'Food Conservation'. The speakers presented papers in which this subject was studied from various angles. Some dealt with the technology of food conservation - how it was done and how the conserved products were used in the diet; others talked about the function of food conservation in particular societies; and nearly all the speakers touched to a greater or lesser extent on the historical aspects of the subject.

This was what we expected and also what we wanted.

For the purpose of this publication, the papers have been divided into three groups, according to what seemed to us to be the main emphasis of each: technology, culture, or history.

We hope that this volume will prove to be of wide interest and that it will serve to encourage further research.

The Editors
February 1988

Salmon fishing. Fra Olaus Magnus: Historia om De Nordinska Folken *(Roma, 1555).*

CONTRIBUTORS

Abdalla, Michael, b.1952. PhD. Lecturer. Poznan, Poland.

Aka, Christine, b.1962. Cand Phil. Volkskundliches Seminar, Münster, West Germany.

Apte, Mahadev, b.1931. PhD. Professor. Duke University, North Carolina, USA.

Apte, Judit, b.1943. PhD. Consultant. Duke University, North Carolina, USA.

Arnott, Margaret Louise, MA. Assistant Professor. Philadelphia College of Pharmacy and Sciences, USA.

Barth, Anke-Sabine, b.1961. Student. University of Münster, West Germany.

Borda, Beatriz, b.1956. BA. Research assistant. University of Lund, Sweden.

Dembińska, Maria, b.1926. Prof D PH. Professor. Warsaw, Poland.

Fenton, Alexander, b.1929. MA, BA, D Litt. Research director. National Museums of Scotland, Edinburgh.

Fjellheim, Bjørn, b.1952. Mag Art. Museum Curator. Floro, Norway.

Gordon, Bert, b.1943. PhD. Professor of History. Mills College, California, USA.

Jobse-van Putten, Jozien, b.1946. Degree Doctorandus. Ethnologist. P.J. Meertens Institut (of the Royal Dutch Academy of Sciences), Amsterdam, Netherlands.

Kisbán, Eszter, b.1936. PhD. Senior Research Fellow. Budapest, Hungary.

Köck, Christoph, b.1962. Cand Phil. Volkundsliches Seminar, Münster, West Germany.

Kowalska-Lewicka, Anna, b.1920. PhD. Krakow, Poland.

Ludvíková, Miroslava, b.1923. PhDr and CSc. Wissenschaftliche Mitarbeiterin. Brno, Czechoslovakia.

March, Lourdes, b.1933. Horticultural degree. Teacher, writer and assessor. Madrid, Spain.

Riddervold, Astri, b.1925. Mag Art ethnologi. Freelance ethnologist. Universitet i Oslo, Norway.

Rios, Alicia, b.1943. Lic Phil and Lic Psych. Teacher, writer and assessor. Madrid, Spain.

Robertson, Una, b.1939. BA. Freelance historian. Edinburgh, Scotland.

Sabban, Françoise, b.1947. Chef de Travaux. L'Ecole des Hautes Etudes en Sciences Sociales, Paris, France.

Schuchat, Molly, b.1928. PhD (Anthropology). Consultant (applied anthropologist), Washington, D.C.

Stoličnà, Rastislava, b.1948. PhD. Research worker, Bratislava, Czechoslovakia.

Theodoratus, Robert J., b.1928. PhD. Professor of Cultural Anthropology, Colorado State University, USA.

Weinhold, Rudolf, b.1925. Dr sc phil. Research Director. Akademie des Wissenschaften des DDR, Berlin, DDR.

van Winter, J.M., b.1927. Dr Litt. University Professor in Medieval History. Utrecht, Netherlands.

OTHER PARTICIPANTS

Benker, Gertrud
Bringeus, Nils-Arvid
Castor, Birgitta
Clark, Jean
Davidson, Alan
Ellison, Audrey
Ericsdotter, Liselotte
Gisladottir, Hallgerdur
Gordon, Mrs B.M.
Jerome, Norge
Lerche, Girth
Lysaght, Patricia
Neubert, Gerd

Nordstrom, Ingrid
Norenberg, Inger
Noklebye, Kirsten Katrine
Rautavuori, Rauni
Ropeid, Andreas
Salomonsson, Anders
Sjogard, Goran
Skjelbred, Ann Helene
Solberg, Karin
Theodoratus, Kay
Uusivirta, Hilkka
Walker, Bruce
Weaver, William Woys

Historical Papers

METHODS OF MEAT AND FISH PRESERVATION IN THE LIGHT OF ARCHAEOLOGICAL AND HISTORICAL SOURCES

Maria Dembińska

The author seeks to show how old some methods of meat and fish conservation are; how and why they were applied; and how they have been modernised. She points out both the differences and the striking similarities between these methods in Poland and other Slav regions; and comments that the latter must have been caused by an identity of needs and of empirical observations.

The greatest worry of primitive man, which probably predated even the knowledge of animal husbandry and of production of agricultural food products, was food preservation and storage. Under primitive living conditions, when winning the food for everyday consumption was a task calling for an enormous effort and was performed, sometimes, at the risk of life, the raw material obtained was too precious to be spoiled. An animal killed with difficulty, a fish, the gathered fruit, grains, or roots could not always be consumed immediately, and required care if they were to be preserved for a long time.

Here, two basic problems emerged; one related to the method of protection of the raw material against rotting, and the other to the method of storage (a storehouse, cellar, a hole in the ground, a pot, a sack, a barrel, etc). Knowledge of how to preserve and to store was a decisive factor for living standards. The mastering of better and better methods of food preservation decreased the greatest danger to humanity - hunger.

Documentation of the history of protection of food raw materials can be found in three kinds of sources: archaeological, written and ethnographic. The archaeological sources supply materials mainly on the oldest methods of food storage and explain certain phenomena related to preservation of food in the distant past, especially when they can be verified through comparison with the sources of other kinds.

The earlier, or later, written sources yield mentions, depending on the particular culture, or epoch under research. These mentions often have to be separated from the general context, often not related at all to the subject we are concerned with, which makes the research extremely difficult.

The ethnographic sources—the most abundant and the ones which most often register cultural relics of village communities—supply very rich verifiable material, permitting comparisons with the mentions in the sources from earlier centuries, or with objects retrieved by archaeology. The storing and preserving technologies based on natural methods have been maintained for an exceptionally long time in folk culture, much longer than many other elements. Some help can also come here from preserving processes used even today by people living out of touch with modern technology.

I shall restrict myself in the present report to preservation and storage of food and food raw materials of animal origin, mainly in Poland or, more broadly, in the Slav regions, but I shall not hesitate to quote examples from the European sphere, or even extra-European. Primitive methods of food preservation are known to people nearly all over the world, irrespective of their interrelations or geographic stituation. It is true that certain forms and customs have been transferred to and adopted by the neighbouring peoples. However, the same need caused the discovery of the same processes, although the inventors were sometimes removed by tens of thousands of kilometres from each other. The cupola oven necessary for baking bread with yeast or leaven, for example, can be found in neolithic cultures in the far Ukraine and in the culture of North American Indians. Similarly, preserving processes were different in details, e.g. in the preservatives used, but the nature of the preserving of the raw material remained the same.

The basic need which led to the preservation of raw materials was, apart from the necessity of protection against hunger, the mass character of these raw materials. Fish and meat were usually obtained through organised fishing and hunting expeditions. Fish, in particular, appearing in big shoals at a known place and time of the year just forced fishery and often the fish had to be transported for a considerable distance after landing. The archaeological sources give us an indirect proof of that, as paleolithic drawings indicate that marine species of fish were eaten in regions distant from the sea; which testifies to their preservation before transportation.[1]

More proof of the necessity to preserve fish comes from the neolithic period (about 5000-2000 BC). Large quantities of fishbones and scales of cod found in neolithic excavations indicate that periodical fishing activities took place, probably in winter, along the Irish and Norwegian coasts. The considerable quantity of slate knives discovered in the same places is to be remarked, since these could have been used for cutting and cleaning the fish.[2] Similar fish remains have been found in neolithic excavations in Poland, for example in Pomerania.[3]

We do not know much about the preservation methods used in these distant epochs. The written sources recording phenomena since about 1000 BC, as well as the later Greek and Roman sources, and later still the medieval sources, together with examples gathered among primitive peoples cut off from the impact of technical civilisation even today, induce us to believe that drying was the oldest method of preservation of fish and meat. The later descriptions and practices enable us to reconstruct roughly how this process was applied in the most distant epochs. The drying was done in the sun and in the open air after previous slicing and cleaning of the fish. This method is used even today in some countries. Cod, herring and pike, having dry meat, were most suitable for preservation through drying.[4] The first written mention of drying products of animal origin can be found in Herodotus. He records in his *Persian Wars*, written about 2400 years ago, that Babylonians and Egyptians dry fish in the sun and the wind and store them dried.

After being cleaned and cut into thin slices, the meat was dried in the sun and the wind. The dried meat was hung high under the roof and protected against pests in various ways. Some peoples first cooked the meat and then dried it on a warm hearth, or in a domed bread-baking oven.

Later written sources, originating from the whole historical period, regularly mention drying of meat and fish.[5] Medieval sources mention dried meat and fish in the whole sphere of European culture. The 'Capitulare de villis', an instruction to the administrators of Charlemagne's domains of the 9th century, makes such a mention.[6] We can also find there passages discussing stores kept in the Emperor's cellars and instruction on how to prepare stocks of dried meat and fish.

Polish sources concerning Polish territories also yield records on dried fish. For example, there is a passage about a fisherman who was marooned by a storm on an uninhabited Baltic island in the 11th or 12th century and lived on dried fish.[7] The Latin term for these fish is *pisces sicci*.[8] The later, 13th and 14th century sources mention *streck-fuss* and *stockfish*.[9] These terms were used to describe fish (usually cod and herring) dried as hard as wood and strung on a stick, usually 30 pieces to a stick. Later, when demand developed, during the 16th century, and when other preservation methods became more popular, dried fish and meat became the food of the poor. They also constituted ship's stores for longer voyages and food for the soldiers during what were sometimes long military expeditions.[10]

There is only one step from drying to smoking of products of animal origin, but we do not know how meat preservation through smoking began. It is believed that the development was incidental, as the dried fish or meat were often hung under the roof of huts, often around the aperture through which the smoke from the fire was escaping. Drying over a smoking hearth could thus have introduced smoking as a form of meat preserving relatively early, although probably unintentionally. Smoking is not mentioned by Herodotus. It is believed that this method of preservation was discovered much earlier (probably together with drying) by peoples living in regions where climatic conditions require heating and drying on the hearth, instead of in the sun and wind as was possible in a milder climate.

We learn already from Greek and Roman sources about consumption of mainly smoked meats in the European region. In Greece, whose culinary and processing skills were known for several centuries BC, various kinds of smoked meats were prepared and classified according to place of their origin. For example, the smoked meats from Alexandria were considered a delicacy.[11] The choice of hams and other smoked meats was even larger in Rome, especially after the conquest of Gaul. This country was famous for its knowledge of meat smoking and supplied mainly smoked pork flitches to its conquerors. Products coming from the lower Mense region later called Westphalia. The tradition of the famous Westphalia hams is, therefore, long. In the Roman Empire, small smoked sausages were made in Lucania (southern Italy) and distributed in the whole of Italy.

The earliest surviving description of smoking meat is furnished by Cato the Elder, living in the second century BC. It is in his work *De Re Rustica* (On Farming). The fact that the description combines three methods of preservation—salting, pickling and smoking—testifies to considerable expertise. He describes the sequence of processes as follows:[12]

> Salting of ham and *ofeliae* [small chunks of pork] according to the Puteoli [a small locality in southern Italy] method. Hams should be salted as follows: in a vat, or in a big pot. After buying hams [legs of pork] cut off the hooves. Use half a modius of ground Roman salt per each ham. Put some salt on the bottom of the vat or pot, and place the ham on it skin downwards and cover it with salt. Then, place the next ham on it and cover it with salt in the same way. Take care that the meat does not touch ... After all hams are placed in this way, cover them with salt so that the meat cannot be seen; make the surface of the salt smooth. After the hams stay for 5 days in the salt, remove them all, each with its own salt. Those that were on top should be placed on the bottom and covered with salt as previously. After twelve days altogether take the hams out; remove the salt, hang them in a draught and cure them for two days. On the third day, take the hams down, clean them with a sponge, smear them with olive oil mixed with vinegar and hang them in the building where you keep the meat. No pest will attack them.

As the recipe comes from the south of Italy, where Greek colonies were situated, it is likely that this method of pickling and smoking was also known in ancient Greece. And it seems that Cato's recipe could be used even today. The use of oil is noteworthy, since oil plays an important role in preserving products, as do other fats.

We do not have any description of smoking fish in the ancient times or in the Middle Ages. Fortunately, the evidence came from the material sources. In the 1960s a Polish archaeologist discovered a settlement in Biskupin (Znin district) of which the estimated date is between the 8th and 10th centuries. It was a production settlement. There were 34 strange pear- or bag-shaped holes and 16 hearths among the relics. Large quantities of fish bones and scales were discovered at the bottom of these holes. Professor Z. Rajewski, D.Sc., the well known archaeologist and researcher who discovered the site—after a very minute analysis with ichthyologists —determined that the relics are probably the remains of a smoking plant for smoking fish caught in Biskupin Lake, for mass consumption. The hearths, with their thick layer of burnt matted particles (earth and refuse, including fish refuse), were probably used for drying the fish, strung on a rod held by two supports, on each side of the hearth. The holes were used for smoking.[13]

The fish bones and scales were analysed ˙y ichthyologists. The fish were mainly pike, bass, roach, bream and very big catfish. The size of these last suggested that they would have had to be cut up for processing; and this was later confirmed by cuts on some of the vertebrae which were found. An attempt at smoking was then made on the basis of the data collected.[14] After scaling and gutting the fish, they were soaked in brine for two hours, and the smoking process was then started. Specialists in the subject of smoke preservation, taking into account the artefacts which had been found, concluded that the Biskupin holes were suitable only for a 'hot smoking' process. The following procedures are involved in this process:

(a) drying of the product surface, the heat denaturation of proteins in the steamed tissue and redrying of the product in the temperature of up to 100°C to remove humidity from the meat - the open hearths over which the fish were hung were used for the purpose;

(b) smoking of the fish, whole or in pieces, hung on a rod in the hole, in temperatures not exceeding 40°C, in the smoke of oak sticks, other wood or green juniper boughs—the holes were covered either with planks or a straw roof, and the process lasted 1½ to 2 hours.

The ability to perform these activities testifies to considerable experience. The product obtained experimentally was of a golden-brown colour and very tasty. The process required constant supervision because when the fish remained in the hole too long—over two hours—they became tasteless, turned tough, burned and lost the necessary fat.

This experiment with the use of the authentic 9th century equipment revealed to us the secret of the smoking technology of those, and probably also much earlier, times. Modernisation of the system led probably to moving the smoking equipment to the ground surface. This can be noticed in cultures where smoking plants for fish and meat, sometimes in the form of big barrels, are situated in gardens next to farm buildings. Such plants already existed on the Hel peninsula in the 15th century.[15] However, we do not know anything about their shape. In spite of the fact that this method of preservation was quite generally known, the process itself was used rather rarely and as an auxiliary method. The written sources (the Polish ones, at least) mention smoking much more rarely than salting or drying. Fish were smoked more frequently than meat in the 13th and 14th centuries. Smoked fish are termed *pisces semiassati* in the sources, while smoked meat is called *carnes fumigatae*. Meat was often treated with smoke in the process of salting and pickling, just as in the description by Cato. The same solution was probably applied by King Ladislaus Jagiellon in the 15th century. According to the chronicler Długosz, he transported the meat from big game 'in the salted and smoked form when warm weather started' and 'required it to be stored' for war purposes.[16]

K. Moszynski, a well-known researcher into the folk culture of Slavs also concludes that smoking itself was not the most popular form of food preservation in the folk culture.[17] The smoking process was largely accidental when dried meat was hung under the eaves, or next to the smoke hole. In Podlesia (Poland), chunks of mutton or pork are kept in brine for a time, hung over a fire to dry, then packed in sewn bags and kept under the roof.[18] This method has certainly been used for a long time in geographically distant localities. So has another method used by American Indians for centuries: thin slices of reindeer, moose or deer meat, or of fish, are dried and smoked over a fire and then crushed with bear fat added (about 1/3 by volume) and tightly packed into leather bags together with dried berries. In a dry place, such meat can be kept for 4 to 5 years.[19] In White Russia, elk meat was cut into chunks, dried in bread ovens and kept in the loft tightly packed in bags.[20] Differences between preservation methods often consisted in the use of various ingredients. For example, a different fat and different preserving spices, or fruit, were used by Indians for making their 'pemmican'.

While drying and intentional or accidental smoking are the oldest methods of preserving products of animal origin, there is more material in the sources on salting or pickling. As soon as the preserving characteristics of salt (marine, from springs, or rock salt) were discovered and the methods of obtaining salt mastered, this method of preserving spread rapidly. We believe that the first salt mines belong to the Bronze Age (about 1800 years BC),[21] but the properties of marine salt were probably known earlier. Herodotus, writing in the 4th century BC, says that Babylonians and Egyptians pickled fish and poultry in salty sea water.[22] He mentions also that the peoples living at the estuaries of the Dneper and the Boh did the same. They salted in the marine 'brine' sturgeon, salmon and catfish, and were thus enabled to eat them raw.[23] We can at once infer that the marine brine was condensed through vaporising as otherwise there would not have been enough salt to preserve the meat. Cato, who lived much later, confirms the inference; he deals in two passages with preparing and making salt and with obtaining salt from sea-water so that 'the solution was suitable for pickling meat, cheese and fish'.[24]

Salt without other additives was used mainly for preserving fish in Mesopotamia, Egypt, Greece, Rome and throughout Europe. Together with the knowledge of the use of salt, salting of various kinds of fish, mainly herring, cod and tuna became widespread. Salted fish are also mentioned in the 'Capitulare de villis' of the time of Charlemagne.[25] Great tuna catches took place for ages along the coast near Marseille. During the 16th century the tuna were still salted according to an old method reportedly originating from Byzantium. The fish were cut into triangles and salted in casks packed tightly in layers.[26]

Information on the salting of herrings on the Baltic coast comes from written sources as early as the 10th century.[27] This required large quantities of salt. Archaeological research has confirmed the existence of salt-making teams near Kołobrzeg as long ago as the early Middle Ages.[28] The autumn herring catches called for rapid preservation. They were packed in barrels, covered with salt and shipped to southern Poland, where they became available on the market from the 11th or 12th centuries.[29] Demand for the product was increasing along with the development of Christianity and the appearance of churches and convents, followed by regulations requiring fasting for about one third of the year. Around the 11th century fish began to be a fasting dish. The sources record that convents situated far from the sea—in Miechów, Lubiaz and Trzebnica, (south-west Poland) were importing casks of herrings from the Baltic region in the 12th century.[30] It probably happened quite often that improperly salted fish got spoiled on the way. Gallus Anonimus, a chronicler of the times of Boleslaus Wrymouth (12th century) mentions that fact in the following lines of a triumphal knights' song in the passage relating to the capture of Kołobrzeg (on the Baltic coast) in 1121:

Pisces salsos et foetentes apportabant alii Palpitantes et recentes nunc apportant filii.[31]

We do not know, of course, whether the word *foetentes* meant a spoiled product, or the unpleasant smell of well-salted herrings. It is generally believed that the Dutch considerably improved the system of herring salting during the late 14th century. They started to clean the fish through their throats and to pack them tightly and very carefully in casks.[32]

Smoking after previous salting of herrings was also introduced in the Netherlands early in the 15th century.[33] It was in this way that those pickling methods appeared which brought an unprecedented wealth to the Netherlands.

Certain relics of the old methods described by Herodotus[34] survived for a long time in the processes used by village people in preserving fish for winter. Small fish, mainly thunder-fish and smelt were thrown in casks, salted and mashed until a nearly dry mass similar to dough was obtained. The fish was eaten either in this form, or in the form of baked cakes. The better the mass was dried, to assume the form of flour or grits, the longer it could be stored. This method was known in different parts of the world. In east Poland (Podlesia) it was still familiar during the 1920s.[35]

In the region of Saratov on the Volga, fish were salted in rectangular holes in the ground walled with planks, as follows: they were gutted and cut into pieces, salted and stacked in layers. There was a cavity to hold a vat to collect the liquid escaping from the fish, so the fish would not be excessively soaked.[36] These old methods probably originate from the very distant past, when pickling of meat products was more popular.

The first mention of pickling in brine comes from Herodotus. Other Greek and Roman writers also describe similar methods while Cato quotes a more sophisticated recipe for brine of sea water with some salt and strong wine added.[37] For more detailed descriptions of the preparation of pickles we have to turn to medieval sources. Several, perhaps even more than a dozen, recipes were known at that time, depending on what was to be pickled and how long the product was to be kept. Pickling liquids were called *salsum*. In Polish, they were often simply called 'salt' probably because salt was the most important component of brine.

The oldest known medieval cookbook dates back to the early 13th century. Two versions survive in Medieval Danish, one in Icelandic and one in Low German. The work contains very interesting descriptions of various pickles of meat and fish.[38] It states how long they last and what they are called, depending on the kind of product being pickled. The version from which I quote is in Medieval Danish, and the names of certain ingredients cannot be explicitly translated. However, I give several examples:

The pickle called *salsum* which can preserve meat for half a year was composed of the following ingredients: salt; cardamon; pepper; ginger; *naghlae* (clove-tree? unknown at present); cinnamon, the quantity of which had to be much greater than that of other condiments; white bread; and a very strong vinegar.

This pickle was used to preserve buck meat (roe deer?) and poultry. The buck meat had to be baked and the poultry fried in a pan. After the product was salted, it was put into the cold brine. The last sentence indicates that the brine was boiled after the ingredients were mixed and cooled before use.

Another kind of pickle seems also to have been used for meat preservation and was sufficient for 3 months storage. Salt was mixed with white mustard seeds (probably ground), 1/3 part of honey, 1/10 aniseed and cinnamon and strong vinegar.

Still another pickle was used for keeping fish for 3 days. It was composed of salt, mint, parsley, cinnamon, pepper and vinegar. The fish for pickling had, according to the recipe, to be cleaned and well doused in oil. Meat and fish could also be pickled for 24 days in brine containing salt, white mustard seeds, 1/4 part of honey and vinegar.

We can see, therefore, the important role that was played by spices, condiments and various seeds in meat preservation; they not only improved the taste, but also had a preserving capacity (e.g. dill, parsley, white mustard seeds and aniseed). Vinegar and mustard prepared from white mustard seeds are the base of the recipes. Methods of preparing these items can be found in various sources.

The 14th century French cookbooks, for example, advise to grind white mustard seeds with olive oil and must. The Latin term *mostum ardum* was probably the source of the word mustard (French moutarde).[39] A similar recipe for preparing mustard was given by Simon Sirennius, a Polish herbalist of the end of the 16th century. In this Polish recipe the must is replaced with vinegar and the olive oil with cooking oil.[40]

Several methods of preparing vinegar also were known. It may be of interest that several pages which survive from a 16th century Polish cookbook quote 9 methods of preparing vinegar.[41]

Honey also played an important role in preserving products. Its ability to preserve had been known for a long time. In the Danish book, we find a description of a meat preservation process similar to a description by Apicius, a Roman gourmet living in the 2nd century AD to whom are customarily attributed the ten books called *De Re Coquinaria*: 'Preserve your meats in mustard mixed with vinegar, salt and honey.'[42]

We should not ignore the use of fats for meat preservation. Cato, for example, recommends smearing smoked meats with olive oil before storing them and Apicius mentions cooking oil as an ingredient of pickle. Our chronicler Gallus Anonimus mentions storing of meats in 'wooden pails',[43] which suggests that it was probably salted and immersed in fat. A passage that can be found in Polish accounts of the 14th century describes this method of meat products preservation in the following way:

'*Triginta pernas carnium pro lardo salsandas*', i.e. '30 chunks of meat to be salted in lard'.[44]

This method, known in the distant past, is still used in villages today. Chunks of meat are immersed in melted salted lard in pots and stored for several months in a cool place.

Finally, we should not ignore still another method of preserving game used in Poland during the Middle Ages. This method was described in detail by a German annalist, Ulrich von Richental[45] early in the 15th century. King Ladislaus Jagiellon, wishing to honour the participants of the Konstanz synod of Roman Catholic bishops at which Poland played such an important role, sent to the Emperor of Germany (Sigismundus) and the Polish bishops the meat of two aurochs, one in 'herring casks', and one was sent whole and unskinned. Its entrails were removed and it was well rubbed with gunpowder and salt and seasoned from inside. The meat protected in this way was transported by cart to Konstanz on the Rhine. It was in winter, as the chronicler says, that the animal was delivered to Konstanz 'on the Tuesday after St Valentine [February 14]'. We do not know whether the meat was fit for consumption; but it probably did not stink, since the Emperor sent it on as a present for the King of England, ordering it to be rubbed again with gunpowder and salt. It was an unusual event and made the Polish king known in an original way. We should remember, however, that the gunpowder used for rubbing the animal played the same role as saltpetre in preserving hams,[46] and any hunter knows well that unskinned game hung by hind legs keeps longest.

NOTES

1. Z. Bukowski, 'Remarks on the fish preservation by the Slav people in the light of archaeological and ethnographic materials ' (in Polish), in *Studia z dziejów gospodarstwa wiejskiego*, vol 9/3, p 53 ff.

2. J.G.D. Clark, *Prehistoric Europe* (Polish edn), Warszawa 1967, p 63.

3. J. Zurek, 'The village from the Stone Age in Rzucevo, distr. Wejherowo, and the Rzucevo Culture' (in Polish), in *Fontes Archeologici Posnanienses*, 1953, vol IV, p 38. See also J. Kostrzewski, W. Chmielewski and K. Jazdzewski, *Prehistory of Poland* (in Polish), 2nd edn, Wrocław, Warszawa-Kraków 1965, p 97, illustration no 22 (map), and p 101.

4. J.G.D. Clark, 'The development of Fishing in Prehistoric Europe', in *The Antiquaries Journal*, An. XXVIII (1948), p 57 ff.

5. See A. Gottschalk, *Histoire de l'Alimentation et de la Gastronomie*, vols I and II, Paris 1948. See also M. Dembinska, *Food Consumption in Medieval Poland* (in Polish with a French summary), Wrocław-Warszawa-Kraków, 1963, p 1.

6. 'Capitulare de villis', M.G. Leges, Sectio II, cap 34.

7. 'Ebbonis vita Ottonis Episcopi Babenbergensis', in *Monumenta Poloniae Historica*, vol II, lib III, cap 4, p 54.
8. *Ibid*, for *siccatorum piscium*. The accounts of King Ladislaus Jagiellon and Queen Hedvige are in *Pomniki dziejowe wieków średnich*, Vol 15 (in Latin), Kraków 1896, p 537 ff.
9. The accounts of King Jagiellon, *op cit*, p 22 ... 'sicci pisces dicti strekfussy'.
10. The oldest accounts of the town of Krakow, Krakow 1877 p 236 from 1392 year: ... 'pro Spytkoni capitanei pro VIII saxagenis de siccis piscibus III marc ...' or from 1461 year: ... 'for the war expedition against Prussia troops ... '1,5 sexageni piscium siccorum'. See also The Accounts of the Governor of Krakow from 1461-71 (ed S. Krzyzanowski), in *Archiwum Komisji Historii Polskiej Akademii Umiejetnosci*, Kraków 1909-13, vol XI.
11. A. Gottschalk, *Histoire de l'Alimentation*, *op cit*, p 103.
12. Marcus Porcius Cato, *De Re Rustica* (ed and translated into Polish S. Los), Wroclaw-Kraków 1956, p 168-9 (cap 162).
13. Z. Kolosowna, 'The early medieval smoke fishery in Biskupin, voyev. Bydgoszcz', in *Sprawozdania Panstwowego Muzeum Archeologicznego* 1950 vol III/1-4, pp 145-52. See also Z. Bukowski, *'Remarks ...'*, *op cit*, for detailed literature on this question.
14. These tests and experiments were done many times by Professor Z. Rajewski, and Professors W. and Z. Szafrański in Biskupin. Two more fish-smoking installations were discovered on the peninsula on which Biskupin is situated; see W. Hensel, *Studies and materials on the History of Early Medieval Great Poland's colonisation* (in Polish), Poznan 1950, vol, 1 p 55 ff, and the same, *The Slav countries in the early Medieval Age* (in Polish), 2nd edn, Warszawa 1965, p 143.
15. Z. Bukowski, *'Remarks ...'*, *op cit*, p 66.
16. J. Dlugossi, *Opera Omnia*, vol V, p 117. This is a Chronicle (in Latin) from the 15th century, edited in Polish by A. Przezdziecki, Kraków 1867 and reprinted by the University of Kraków in 1970.
17. K. Moszyński, *The Folkloric Slav Culture*, 2nd edn, vol I (in Polish), Warszawa 1967, p 236.
18. *Ibid*, p 238.
19. S. Bleeker, *Indians*, New York 1969, p 18. See also *The Basic Everyday Encyclopedia*, New York 1954, p 404 under 'pemmican'.
20. K. Moszyński, *The Folkloric Slav Culture*, *op cit*, p 239.
21. H. Burchard, 'The exploration of salt in Poland before the 13th century' (in Polish), in *Slavia Antiqua*, vol VI, p 399.
22. Herodotus, *The Persian Wars*, p 152.
23. *Ibid*, p 293.
24. Cato, *De Re Rustica*, *op cit*, p 106, cap 83, 'Domestic salt production' and p 115, cap 106, 'The preservation of sea water'.
25. 'Capitulare de villis', *op cit*, cap 34.

26. Archestrates, an Epicurean poet of the 4th century BC, noted this kind of preservation. However, we only know his recipes at second hand, from the work of Atenes, who lived in the 2nd century and noted them from the books now lost. See A. Gottschalk, *Histoire de l'Alimentation*, *op cit*, 'Textes justificatifs et Addenda', p 358.

27. *Thietmari Merseburgensis Episcopi Chronicon*(ed Z. Jedlicki, in Latin and Polish), Poznan, lib IV, cap 45, p 209.

28. L. Leciejewicz, 'Early medieval Colberg' (in Polish) in *Slavia Antiqua* 1960, vol 7, p 349 and map 7.

29. Noted in the sources from the 12th century. See *Cod. diplomaticus Poloniae Minor*, vol II, no 375, on customs duty for herrings. See also Cod. Diplomaticus Masoviae (ed S. Kochanowski), nos 88 and 469.

30. K. Maleczyński, The oldest markets in Poland and their relation with the towns before the colonisation on the so called German law (in Polish), in Studia mad Historiq prawa polskiego, vol X, fasc 1, Lwow 1926 p 31 ff. See also *Cod., Pommeraniae*, no 99, p 232.

31. 'Galli Chronicon', in *Monumenta Poloniae Historica*, (ed A. Bielowski) 1864, vol I, lib II, cap 28, p 447.

32. R.J. Forbes, 'Food and Drink', in *A History of Technology*, vol II, Oxford 1956, p 123-24.

33. A Gottschalk, *Histoire de l'Alimentation*, vol II, p 299.

34. Herodotus, *op cit*, vol 1, p 112.

35. Z. Bukowski, '*Remarks ...*', *op cit*, p 74.

36. K. Moszyński, *The Folkloric Slav Culture*, *op cit*, p 241.

37. Cato, *De Re Rustica*, *op cit*, p 115.

38. The manuscripts of two small cookbooks, written by H. Harpestraeng in the 13th century can be found in the Royal Danish Library, edited in Medieval Danish by M. Kristensen, Copenhagen 1908-1920. Dr. Rudolf Grewe has published an important essay on these and the corresponding Icelandic and Low German manuscripts in *Proceedings of the Cambridge Symposium on Current Research in Culinary History*, Culinary Historians of Boston, Cambridge, Massachusetts, 1986, pp 27-45.

39. J. Muszyński, *Vegetables, fruits and condiments in Poland in the 14th century* (in Polish), Warszawa 1924, p 21.

40. D.S. Syrennius, First books about all kind of herbs, - antique printed book from the end of the 16th century, cap 1202-3, 1613, University Library in Warsaw, sign no 5.

41. Cookbook, rest of a Polish ancient printed book from the first part of the 16th century, (ed Z. Wolski), Biała Radziwiłłowska 1891.

42. *Les dix livres de cuisine d'Apicius*, traduit du Latin avec commentaire par B. Guégan, Paris 1933, book I, cap 9, pp 15-16.

43. *Galli Chronicon*, *op cit*, lib I, cap 2.

44. The accounts of King Jagiellon and Queen Hedvige *op cit* (see note 8), pp 129, 565.
45. *Chronicon* of Ulrich von Richental from 1417, The Tsar Archaeological Society, Petersburg 1874. (This Chronicle had been first published in Old German in Augsburg between 1420 and 1423.)
46. For the preservation of meat, potassium nitrate (KNO_3), (also called Indian saltpetre) must be used, not sodium nitrate ($NaNO_3$).

FASCISM, THE NEO-RIGHT, AND THE PROBLEM OF GASTRONOMY

Bertram M. Gordon

The paper delivered by Professor Gordon, briefly described below, was subsequently revised and expanded in some respects and presented a second time at the Oxford Symposium on Food History later in June 1987. Since the full text of the revised version is published in *Taste: Oxford Symposium Documents, 1987* the text of the Sogndal version is not included in the present volume.

The paper explores the nature of fascism and discusses to what extent the various forms of fascist theory had implications, explicit or implicit, for gastronomy. Italian fascist theory is considered, including the views of the Futurist leader, Marinetti. Paintings executed in Italy during the fascist period are scrutinised for signs of these influences.

The second case studied, that of Nazi Germany, is shown to be of a 'tighter' nature than the Italian one. Hitler's personal attitudes to food and eating habits are discussed. There is an account of the significance of Eintopf (one pot) cooking. Paintings, again, are examined for the evidence they may afford.

Further material from France, Spain and Portugal is discussed. The difficulty of following fascist theory on gastronomy into wartime, whether the Spanish Civil War or the Second World War, is explained at the end of the paper.

'MAY HIS PIG FAT BE THICK': Domestic Conservation of Fat in Hungary

Eszter Kisbán

In the pre-industrial period different methods of conservation prevailed in different parts of Hungary, depending on the ecological and economic systems of the various areas. The most important item of conservation was lard, whose price was always higher than that of meat. Lard was present in all wages paid in kind and was used in both peasant and upper-class kitchens. In Hungarian peasant culture, it was the main food of cold meals.

It has been customary, since the Middle Ages, to describe Hungary as a country with a wealth of foodstuffs. Indeed, in the 16th century the Hungarian economy depended on the export of live animals, chiefly cattle. These accounted for 93.6% of exports, while other foodstuffs (wine, honey, fish) accounted for another 4%. At this time, food constituted only 8.7% of imports (mainly spices, wine, herring and salt).[1] It is of interest that the driving of cattle to the markets of Vienna, the cities of South Germany, and Venice continued to be the main feature of Hungarian food exports until the middle of the 19th century, when wheat assumed first place.

The abundance of foodstuffs was accompanied, as one would expect, by many traditional methods of preserving them. In general, these followed the pattern familiar in other parts of Europe. However, the preservation of animal products, and above all that of pork fat, has had a special importance, and it is this subject which is explored in the present paper.

ANIMAL RESOURCES IN THE TWO MAIN REGIONS

In considering the animal resources of Hungary, and the uses made of them, it is necessary first to draw a distinction between the lowland regions and the Carpathian Basin.

The famous 'big cattle' of Lowland Hungary and the smaller stock of other areas of the country were mostly eaten fresh at home. In neither case was the regular conservation of their beef known (at least not since the 16th century - a report of the 14th century mentions that Hungarian soldiers in Italy carried dried, powdered beef). In more recent times, herdsmen in the Lowlands who spent all their time at the grazing areas, made their own dried meat (beef as well as mutton) by stewing it in small pieces and drying it in the open. This was eaten as it was, or put into water and made into a soup.[2]

The Carpathian Basin with its ranges of hills, was a zone of oak forest, which provided mast for pigs. Maize, the best pig-fodder, became acclimatised here from the 17th century onwards, but only at low altitudes. The main form of animal husbandry was keeping sheep for milk. So it was the sheep in the mountains (and pigs lower down) which were the main animals slaughtered for domestic use. The fattening of pigs for household consumption, using bought fodder, spread to the mountains only in the 19th century.

The year's supply of meat preserved by salting and smoking was mainly of sheep in the mountains, and exclusively of pigs in the Basin. The differences in the type of animal had other consequences also. For cooking and for cold collations, sheep's milk products were used in the one area and lard in the other. The main sheep's milk product was cheese made with rennet. This was then kneaded with salt and put in a sealed tub, in which it could keep for years. We shall return later to the question of lard.

Beyond the sheep-farming area, milk products, fresh or preserved, were made from cow's milk. A soft curd was made from sour milk. Some of this was preserved for Lent and Advent by kneading it with salt and shaping it into small, flat cakes, which were then dried or smoked. Until the early 20th century, during Lent and on certain week-days of Advent, Roman Catholic peasant households cooked with butter, and in some regions also with oil, instead of with pig's fat. They accumulated butter for this purpose over a long period. After every churning the butter was rendered to get rid of the remaining buttermilk.[3] The rendering of butter, without salting, characterised all of southern Middle Europe. Here, butter was used for cooking, and its spreading on bread played no part in the traditional food culture.

FAT: THE MOST IMPORTANT ELEMENT

Amongst all the Hungarian foodstuffs, from the Middle Ages to the early 20th century, conserved pig fat played a quite outstanding role, except in the Carpathians. In the past, people valued foodstuffs above all for their energy content, so it is little wonder that peasants and agricultural workers gave pig fat the leading place. It was part of every payment of wages in kind in the 17th as in the 20th century. A plentiful supply was a sign of prosperity, as symbolised by numerous sayings and in the people's way of thinking, even in the very recent past. 'May his pig fat be thick!' is a wish that someone should succeed in his business.

In this paper, we use the term 'pig fat' to indicate big pieces of fat, in conserved form, salted, and dried or smoked; we reserve the term 'lard' for fat, straight from the pig, that has been melted down. For as long as records exist, from early post-medieval times till about 1900, conserved pig fat was generally more costly than any kind of meat. This was no longer so in Budapest by 1900, but it remained true in the towns of the countryside.[4]

Pig fat played such an important role in the system of meals and in cooking throughout Hungary, we can almost say that Hungary thereby had a characteristic food system. The situation was not unique in Middle Europe, but was far from being general.

The earliest Hungarian food statistics, referring to 1884, and compiled by Keleti, provide the per capita consumption of the total population, and additionally, that of the city-dwellers. To indicate the relative proportions, we give below the data referring to pig fat, lard, butter and meat. The consumption of edible oil and margarine was then so low that these statistics ignore them.[5]

AVERAGE PER CAPITA CONSUMPTION OF EDIBLE FATS AND
MEAT IN HUNGARY IN 1884

	in the Whole Population	in the Urban Population
Fat	18.56 kg	25.59 kg
Lard	8.1	11.13
Butter	1.24	3.07
Meat	32.93	62.19

The consumption of pig fat in the cities was therefore greater than in the villages, because the cities were on the one hand strongly agrarian and, on the other, had higher living standards. The statistics themselves clearly indicate the prestige fat had, where its higher consumption was one form of reflection of higher living standards.

Contemporary statistics relating to these headings are not easy to find elsewhere, but Teuteberg's data for Germany in the same year, 1884, show an annual per capita consumption of fresh pig's fat, together with smoked ham, sausages and conserved pig's fat amounting to 16.29 kg, which is less than the figure for pig's fat alone in Hungary.[6]

Behind the averaged out figure of 18.56 kg lay a central area of Hungary where in some counties consumption varied from 26 to 31.5 kg a head. In the Southeast, in Transylvania, averages of 20 to 23 kg were characteristic. In the highest consumption areas, agricultural labourers employed by the year got 20-60 kg of pig's fat as part of their wages, which was additional to that from the pig killed in their family. Seasonal migrant workers who were employed at peak periods in the summer months got, and consumed on the spot, 4-5 kg of fat a month.[7] At their homes, at least one pig a year was also fattened.

The Keleti statistics caught the phase in a time when fat consumption was certainly at its highest. The data-gathering was done after the introduction in the 19th century of a new type of pig, the *mangalica*, which was bigger and more productive in fat by live weight (48.16%) than the old type (39.05%), an improvement of 10%.[8] If two *mangalica* pigs, each with a live weight of 230 kg, were killed, a family of five would get an average amount of 30 kg of pig's fat, as well as 11 kg of lard, each.

Only one attempt has been made by Hungarian economic historians to calculate food consumption, and that only for the 17th century. On the assumption of one small pig slaughtered per family, they arrived at an annual figure of 11 kg of pig's fat per capita.[9] But it is likely that more than one small pig was slaughtered, and consumption may easily have been about 20 kg a head. In the 17th century nearly all the pig's fat was conserved because the custom of melting down part of it to make lard was not practised then; when a pig was slaughtered, only small amounts of intestinal fat were rendered.

Modern statistics do not, unfortunately, run continuously, and it is necessary to select years when figures relating to fat and lard were separately available. The proportion of fat to lard in 1884 was 2.29 : 1. By 1956-58, the proportion was reversed, with a fat to lard consumption ratio of 0.22 : 1, i.e. 3.9 kg of fat and 17.06 kg of lard per capita.[10] I do not intend to analyse the components of this reversal here, but it should be made clear that the regionally characteristic food-system marked by considerable use of pig's fat is a matter of history, the last traces of which are now fading away.

In both civil and military inventories of the early post-medieval period, there is much reference to pig's fat, reckoned either in complete pieces or in halves. They contain references to only small quantities of rendered butter in wooden containers, and to even smaller amounts of oil. They never mention lard, however.

In the 20th century, in the eastern part of present-day Hungary and in Transylvania, examples can still be seen of pig's fat in complete pieces.

To get these, the pig was opened along its stomach, and the fat around the whole animal was stripped out as a unit, from which nothing was removed. Not even the edges were trimmed. It was conserved complete. If half pieces were wanted, the fat was cut longitudinally, and in more recent times the edges were trimmed. Throughout the country, conservation was by dry-salting, after which one of two traditional processes could be applied: air-drying only in the Great Plain, and smoking everywhere else.

We do not know how this difference arose. The complete units of fat to be dried or smoked were so wide that two rods of wood were needed to keep them flat, so that they should not curl up in the process.

Until the end of the 17th century, no one rendered fat at the time of the slaughtering and so neither peasant nor upper-class households stored lard. This is made clear from a new cookery book published in 1695.[11] It was only in the 18th century that upper-class households changed from cooking with fat to cooking with lard, and peasant households slowly adopted the practice also. From this time on, a division was made at slaughtering between pig's fat to be conserved, and fat to be rendered into lard. Peasant households in the western areas were more lard orientated, and those in the eastern and central areas preferred fat. The upper and middle-classes turned away from using fat. As against salted or smoked fat, lard has a wider variety of uses, because its taste is neutral.[12]

What kinds of cooking was fat used for, while lard was not in regular use? According to the 1695 cookery book, upper and middle-classes used fat for cooking and baking all kinds of lean meat, vegetables, pulses and gravies. If a roux was required, the flour was fried in fat. With salads, pieces of roast fat were put on top. For all kinds of pastries or noodles, butter was used, because the fat was salty or smoked. Peasant households cooked with fat in a similar way, but they consumed most of the fat as a food in itself at meals which did not exist amongst the upper-classes, or else had a different composition.

In the Central Hungarian lowland area there was, right up till the 20th century, a range of types of soup, gruel and meat dishes amongst the peasants and agricultural labourers, the preparation of which began by melting pieces of fat; the food was cooked with the fat, and the crackling, that had been taken from the fat, was laid on top. More important than this regional characteristic, however, was the significance of fat in supplying energy, and in the organisation of meals amongst these people, in a way not characteristic of the upper-classes.

Whilst the upper-classes in Hungary ate according to the medieval two-meal system until about 1700, having hot dishes at each meal, peasants doing outdoor work from spring till autumn had three main daily meals, morning, midday and evening. Two of these were of hot dishes, but one, most likely in the morning, or at midday according to work circumstances, was cold. This arrangement applied to most of the country, except for a crescent-shaped zone near the frontier.

About 1700, the upper-classes went over to the modern three-meal system. The new additional meal was a breakfast of coffee and pastry. The use of morning coffee did not begin amongst the peasants until around 1900 and even then was adopted slowly and reluctantly; the peasants found it incompatible with fat, which they did not want to give up at breakfast.[13]

Breakfast was not the only fat-based meal. The peasants themselves took an afternoon snack on the days of heaviest work, and on the longest days during the summer. Day-labourers, however, insisted on such a snack right through from spring till autumn. In areas of high fat consumption, it consisted of bread with fat. In cool weather in winter, spring and autumn, meals eaten in the fields could be of fried, not cold, fat.

By the first half of the 20th century, there was amongst the agricultural population a depression, stemming from bad economic circumstances and from a developed stage in the subdivision of property holdings between male children. The per capita consumption of fat also fell significantly. There were millions of hungry country folk with a small stake in the land, who were in a constant state of anxiety about whether or not they could get work as agricultural labourers to earn the bread they needed for survival and the fat required to give them enough energy to work.

After the repartition of land in 1945, such people became peasants with enough land to live on and many of them, thanks to increased opportunities for social mobility, graduated into higher spheres of management or moved into towns. But, whatever became of them, there still remained amongst them in the 1970s—even though it had become inappropriate—the concept of the use of fat as the symbol of a secure, wealthy life. There should be fat enough in the larder for people to eat their fill of it. In the words of the saying, for their business to be successful, their fat should be thick.

NOTES

1. *Mágyarország története* 1985, p 306.
2. Paládi-Kovács 1979.
3. Keszi-Kovács 1969.
4. *Magyar Statisztikai Évkönyv*, Uj folyam 8(1900).
5. Keleti 1887, pp 68, 116.
6. Teuteberg 1986, pp 238, 269.
7. *Mezógazdasági munkabérek* 1901.
8. Tormay 1896, pp 218-35.
9. Makkay 1979, pp 260-3.
10. Sós 1959, p 40.
11. *Szakáts mesterségnek könyvetskéje* 1695.
12. Kisbán 1974.
13. Kisbán 1987.

REFERENCES

Keleti, Karl 1887: *Die Ernährungs-Statistik der Bevölkerung Ungarns*, Budapest.
Keszi-Kovács, László 1969: 'Die traditionelle Milchwirtschaft bei den Ungarn', in László Földes (ed) *Viehwirtschaft und Hirtenkultur,* Budapest: pp 640-95.
Kisbán, Eszter 1974: 'Vom Speck zum Schmalz in der ländlichen ungarischen Speisekultur', in *In memoriam António Jorge Dias,* Vol 2, Lisboa: pp 283-96.
Kisbán, Eszter 1987: 'Coffee Shouldn't Hurt. The Introduction of Coffee to Hungary', in *Wandel der Volkskultur in Europa* (Festschrift fur Gunter Wiegelmann), Münster (forthcoming publication).

32   Kisbán

Magyarország története (The History of Hungary), 1985: Pach Zsigmond Pál (ed),
    Vol 3, Budapest.

Magyar Statisztikai Évkönyv, Uj folyam 8(1900), Budapest.

Makkay, László 1979: 'A magyarországi mezőgazdaság termelési és fogyasztási
    struktúrája a XVII század közepén' (The structure of agricultural production and
    consumption in Hungary at the mid-17th century), in Peter Gunst (ed),
    Mezőgazdaság, agrártudomány, agrártörtenet, Budapest: pp 253-263.

Mezőgazdásagi munkabérek 1901: *'Mezőgazdasági munkabérek Magyarországon
    1899-ben'* (Agricultural wages in Hungary in 1899), Budapest.

Paládi-Kovács, Attila 1979: 'Dörrfleisch - Eine archaische
    Fleischkonservierungsweise der Ungarn', in *Acta Ethnographica* 28, Budapest:
    pp 191-204.

Sós József 1959: *Nepélelmezés* (Alimentation), Budapest.

Szakáts mestérsegnek könyvetskéje 1695: (The booklet of the art of cookery).
    Tótfalusi Kis Miklós Publ. Kolozsvár.

Teuteberg, Hans Jürgen 1986: 'Der Verzehr von Lebensmitteln in Deutschland pro
    Kopf und Jahr seit Beginn der Industrialisierung (1850-1975)', in Hans Jürgen
    Teuteberg - Günter Wiegelmann, *Unsere tägliche Kost,* Münster: pp 225-79.

Tormay, Béla 1896: 'Sertéstenyésztés' (Pig breeding), in *Magyarország
    földmivelése 1896,* Kiadja a Földmivelésügyi m. kir. minister, Budapest: pp
    209-68.

THE PICKLING OF VEGETABLES IN TRADITIONAL POLISH PEASANT CULTURE

Anna Kowalska-Lewicka

When Polish peasant diet contained only minimal quantities of meat and consisted mainly of vegetables, flour and buckwheat, it was not surprising that pickling was much favoured, for it improved the taste of the food. The flavours thus enhanced have remained popular to this day, even though the old methods have changed.

By 'pickling' we understand a process of preserving plants for human and animal consumption[1] by generating lactic fermentation in them.Lactic acid delays the growth of microflora and arrests putrefaction. Successful pickling depends on the proportion of soluble sugars in the preserved vegetables; on their moisture; on the temperature in the first few days (15-20° C); and on the exclusion of air. Hence the need for pounding the mass to release juices and air bubbles. Vegetables which are to remain unbroken (e.g. cucumbers) are covered by salt water. Salt is advantageous in the process of pickling but not essential. Aromatic herbs are often added as they, together with by-products of fermentation (alcohol 0.5-1%), improve the flavour.

Pickling and marinating are some of the oldest methods of preserving food, known already in pre-historic times. It developed principally in cool and moderate climates. The drying of food for storage was not practicable in cool regions because of low temperatures and scarcity of sunny days, but storage was essential in view of short summers and long winters.

Pickling was so important to Poles and their Eastern neighbours that some e.g. Lithuanians worshipped a god of pickled food called Roguszys.[2]

Historical research has established that in Poland many wild and cultivated plants, and also flour made from them, were pickled. The preserving of milk, butter-milk and whey was common too but this is outside the scope of this article. We shall also omit the description of plants pickled only sporadically and shall deal only with those that were economically and culturally important and which survived in the peasant culture until the 19th or even 20th century.

It was customary until the second half of the 19th century to pickle the leaves and stalks of a wild, umbelliferous plant (genus *Heracleum*) known in Polish as *barszcz*. Among vegetables the most important were cabbage, white cabbage, and the root of red beet. Until the early 20th century roots of rape, turnip and carrot were sporadically pickled. Cucumbers, apples and rarely pears were also pickled as was the flour of oats, rye, buckwheat and earlier millet. A sour drink was produced from fermented bread.

We shall now discuss more fully various kinds of produce in the order of their importance in peasant cooking (and to a certain extent in the whole country).

Pickled cabbage (sauerkraut) was undoubtedly the most popular among peasants, landed gentry and urban middle classes alike. Cabbage was pickled either whole[3] or finely shredded with a knife or on a 'mandoline' board which became popular in villages at the turn of the present century. Up to the same time it was customary to pickle cabbages in special ditches, which had their sides covered by wooden planks.[4] Layers of whole cabbages alternated with layers of shredded cabbage leaves, the top layer being always shredded. Cabbage was also pickled in precisely the same way in barrels.

Towards the turn of the century pickling of shredded leaves predominated but a few whole cabbages were always included in a barrel to provide whole leaves necessary in the preparation of a dish known as *golabki*.[5] In some districts when whole cabbages were to be pickled they were first scalded with boiling water or quickly heated on a bonfire or in an oven. The cabbages, arranged in barrels or ditches, were next pounded with feet or wooden clubs to release juices and air bubbles to prevent rotting. The cabbage was then covered with a piece of linen cloth and a wooden lid on which rested a heavy stone. At intervals mould was removed from the top of the barrel, the linen cloth rinsed and the wooden lid scrubbed; water was topped up as required. Barrels with freshly pounded cabbage were kept for about a fortnight in a warm kitchen or stable and later transferred to a cellar or larder where they remained for the whole winter.

Apples and, less frequently, pears were sometimes added to the cabbage. Sour and hard fruit from uncultivated trees otherwise unsuitable for human consumption, was used for this purpose. Pickled apples could be eaten as a separate dish or added as flavouring to other dishes.

Cabbage was pickled either with or without the addition of salt, depending to a large extent on the financial status of the family concerned with the pickling. Caraway seeds and dill were added to improve flavour. In some parts of the country oak and cherry leaves were used to improve the preserve. The standard of living in the villages improved in the 20th century and the culinary demands rose. Pickled cabbage is now flavoured with onions and garlic; red beetroot is added to enhance the appearance.

Until recently, an average family pickled several large barrels of cabbage. The pickling, performed in late autumn after the potatoes had been dug, was a communal activity to which neighbours and relatives were invited. Most of the work was done by women, only the pounding being done by men. At the completion of the work there was a reception and often dancing.

In the 19th century and less frequently in the 20th century small quantities of finely chopped roots of rape, turnip and carrot were preserved by methods similar to those applied to cabbage.

Sauerkraut (pickled cabbage) was and still is extremely important in Polish cooking, especially in the villages It is cooked as a vegetable to be served with meat dishes or with dumplings. The liquid dredged from barrels is used as 'stock' for a soup made with the addition of cured pork, lard, flour and flavourings. In Lent and on other fast days meat was replaced by vegetable fats and mushrooms. This soup was so highly regarded that frequently the cabbage itself was thrown away once the liquid was used up.[6]

Pickled cabbage is the basis of a national Polish dish - *bigos*.[7] This is a traditional dish of the landed gentry, frequently prepared for winter game hunting, when a vat of *bigos* was brought to a clearing in the forest and reheated on a bonfire. In the villages *bigos* is served as a festive dish at wealthy weddings. The basis of *bigos* is pickled cabbage to which are added several kinds of meat, bacon, pickled plums and other fruit. The dish is prepared over several days alternately heating and cooling it.

These are the principal cabbage dishes. There are many more; e.g. it is used as a filling for dumplings (ravioli).

In present day Polish menus all traditional cabbage dishes are still represented, albeit less frequently than in the first half of the 20th century. Domestic pickling has been abandoned in towns mainly because of lack of storage space in new houses, but pickled cabbage can be bought at all greengrocers. There are now new ways of serving it both in towns and in villages. Mixed salads of raw pickled cabbage with onion, apples and carrots are popular. A simplified form of *bigos*, with fewer varieties of meat but with the addition of tomatoes, is the main item on the menu in cheap snack-bars and canteens. *Bigos* made according to traditional recipes is considered a formal dish to be served during the winter carnival season.

The method of pickling cabbage has been dealt with in more detail because of its special place in Polish diet in the past and to the present day not only in villages but generally in the whole of the country.

Next in importance in peasant cooking comes a sour soup known as *zur*. A 17th century writer mentions *zur* as a Polish national dish.[8] In every urban or village household there was a clay pot for the fermenting of *zur*. This pot was not generally washed after each use so that a little of the solution was left to facilitate the fermentation. Boiled tepid water was poured over flour in this pot (oats flour in the South, buckwheat flour in the East and North-East and rye flour in the rest of the country). The pot was left overnight at the edge of the kitchen range.[9] In the morning, the water, now called *zur* was sufficiently sour. A soup, also called *zur*, was made out of this liquid. It was made festive by the addition of meat, sausage, bacon, cream and flavoured with garlic, bay leaf and pepper. As a typical soup for days of fasting it had no additions and was only flavoured with salt and garlic.

We know from literature that *zur* was a daily dish throughout the year among all social classes in town and country. Nevertheless it is associated mainly with fasting, especially Lent when it was used instead of fats and milk. In Polish folklore *zur* became a synonym of fasting and self-denial. To celebrate the end of fasting *zur* was 'killed' (as a symbol of fasting and of winter) by burying the *zur*-pot, breaking it or hanging it from a dead branch etc.

After the Second World War *zur* lost its popularity in peasant cooking. It is also no longer a typical fasting dish, but it is frequently served commercially. Factory-produced rye *zur* is sold in bottles, and synthetic *zur* in powder form. It is rarely home-made and then only in villages.

Kisiel is a dish with a long history and it is related to *zur*. It is especially popular in Eastern Poland along the Polish-Lithuanian and Polish-Ukrainian borders. *Kisiel* was prepared in the same way as *zur* but with a larger proportion of flour. The sour liquid was strained and after boiling and cooling had the consistency of jelly. It was made almost exclusively from oat flour.

'White kisiel' was made of flour and water and, apart from daily use, was also served as a traditional dish at Christmas Eve supper.

'Red kisiel' is ordinary oat *kisiel* coloured by the addition of red berries. It was considered tastier. In marshy regions where the red berries grow they were gathered and stored in small barrels.

In exceptional cases *kisiel* was sweetened with honey. Nowadays *kisiel* is only prepared in villages in the traditional way for Christmas, but it has become a popular pudding throughout the country, especially in snack-bars and canteens. It is commercially produced in powder form consisting of cornflour, sugar, flavouring and colouring matter.

It is not possible to deal in the space of one article with all aspects of all pickled vegetables, methods of pickling, storing, their application in the preparation of dishes and their importance now and in the past in the traditional diet of the Polish countryside. We must however mention the always popular sour soup *barszcz*. Up to the 19th century it was made from a wild plant (*Heracleum sphondylium*), which in the second half of the 19th century was replaced by pickled red beet. This soup took over the name *barszcz*.

Pickled cucumbers have always been extremely popular both in towns and in villages. Cucumbers are arranged in wooden barrels and covered with salty water to which dill, garlic and horseradish are added together with oak and cherry leaves containing tannin. In towns cucumbers are now mostly pickled in glass jars rather than in barrels. They are also easily obtained in shops.

Pickling of mushrooms deserves a fuller description. Almost all varieties of edible mushrooms were pickled in the villages, *Lactarius deliciosus* being considered the best. In the 17th and 18th centuries 'a barrel of pickled mushrooms' is often listed among the gifts offered by peasants to the landowner. Mushrooms were pickled similarly to cabbage, and were highly regarded as an addition to warm dishes or bread. They are now rarely encountered due to the scarcity of mushrooms in the woods.

During hay-making and harvesting sour drinks were extremely popular. To mention a few: sour milk, butter-milk, whey, juice of pickled cabbage and cucumbers and the well known *oskola*. This last one is juice tapped from birches, rarely from other trees, taken in early spring, before the leaves appear. Kept in barrels, it ferments and due to the sugar content it has a higher percentage of alcohol (up to 2%) than vegetable juices. It is a great favourite in a heatwave.

A Russian drink made of fermented bread steeped in water and called *kwas chlebowy* became popular in the part of Poland which was occupied by Russia from the end of the 18th century until 1914.

Pickling of vegetables, fruit, mushrooms and flour served not only to preserve the food. It was also used to enhance the taste. In the summer, peasants frequently pickled small quantities of vegetables for immediate consumption. The pickling was complete in 2-3 days and the tart taste was highly appreciated. Even potatoes were thus pickled. This is understandable when one remembers that the peasant diet contained only minimal quantities of meat and consisted almost exclusively of vegetables, flour and buckwheat—all with a bland taste. Foreign herbs and condiments popular in richer households were inaccessible to the poor peasants. Tartness was the only definite taste obtainable from vegetables.

In spite of the addition of new produce to village cooking, and higher cooking skills, sour dishes are still popular, though there is a tendency to replace old pickling by marinating in vinegar or adding vinegar to dishes formerly prepared by fermentation.

NOTES

1. Traditional preparation of fermented food for animals is still practised only in the Carpatian region. Until the middle of this century leaves of turnips, beet, cabbage and others were fermented in huge tubs, and used as winter fodder.

2. Lithuanian Chronicles. Quoted in Linde, *Dictionary of the Polish Language,* 1855, vol 2, p 359.

3. Mainly in regions where a variety of loosely bound cabbage was grown.

4. Fermentation in ditches persisted longest (till the middle of this century) in Southern Poland (Podkarpacie) and in some central regions.

5. *Golabki* (singular *golabek*) means 'pigeon'. It is a traditional Polish dish of cabbage leaves with buckwheat and meat (mushrooms replacing meat on fasting days).

6. See A. Maurizio, *Vegetarian nourishment and agriculture through the ages,* Warsaw, 1926, p 43. Areas where sour soup was a traditional Christmas Eve dish are shown in the *Polish Ethnographic Atlas,* no 7, map 380.

7. Recipes for preparing *bigos* and cirumstances when it was served are described frequently in literature of the 19th century. A detailed description is in the epic poem *Pan Tadeusz* by the greatest Polish romantic poet Mickiewicz. See also Z. Gloger, *Old Polish Encyclopedia,* vol 1, Warsaw, 1972, pp 172-3.

8. Z. Gloger: *Old Polish Encyclopedia,* vol 4, Warsaw, 1972, pp 522-3.

9. J. Bohdanowicz on 'Nourishment' in publications of the *Polish Ethnographic Atlas: Nourishment of the peasant population,* Krakow, 1973, pp 59-60; and *Polish Ethnographic Atlas,* no 7, map 381, and no 5, map 266.

TRADITIONAL METHODS OF FOOD PRESERVING AMONG THE BULGARIANS

Lilija Radeva

The preserving methods described are the most widespread processes for preserving animal or vegetable foodstuffs among the Bulgarians. In the past they played an important nutritional role.

The main methods for the preserving of fruit, vegetables and herbs are drying, boiling, baking, dousing in water or fruit-juice with the addition of natural preservatives.

Meat products (meat, bacon fat and fish) are preserved by salting, boiling and bottling in fat, drying in the air, smoking and pulping.

There is also a variety of methods of preserving milk: the production of yoghurt, the preserving of raw milk by natural fermentation, and cheese-making.

The preserving of foodstuffs of vegetable and animal origin is a very ancient and widespread process in traditional Bulgarian food. Some preserving methods are mentioned by chance, although in a cursory and incomplete manner, in articles and sources by earlier writers—novelists, historians, ethnographers, economists—working at the end of the 19th century and the beginning of the 20th. These include L. Karavelóv, G. Rakovski, K. Jireček, D. Marinov, Iv. Sakăzov, G. Kačarov, V. Decev, J. Zachariev and A. Zlatarov. In the last thirty to forty years complex researches have resulted in the gathering of a large body of source material on food and nutrition which has been published in collections and monographs (Chr. Vakarelski, A. Primovski, G. Krăsteva and L. Radeva). Included in these researches is data on preserving methods, although a complete and systematic specialist study of this question has not been undertaken.

The earliest documentary sources about the food of the Slavic peoples immediately following the formation of the Bulgarian Empire in 681 AD mention that after their migration south of the Danube the Slavs (one of the ethnic components of the Bulgarian nation) gathered wild fruits - pears, apples and plums - and dried them to prepare food from them.

The main diet of the other ethnic element, the proto-Bulgars, on their migration to the Balkan peninsular was horse and sheep meat preserved by drying, prepared using a very primitive process: the pieces of meat were laid under the horses' saddles where they became salty and dried. Then they were eaten raw on campaigns, in battle and in times of peace.

This technique—drying (either fruit or meat)—continued for centuries as one of the oldest and most widespread methods of preserving, but in the course of time it changed and became more varied, and other more complicated methods were added to it.

Also customary was the preserving of milk and the making of long-lasting milk products.

It cannot be exactly determined when new methods of preserving originated, but it is apparent that in the so-called 'traditional period' (from the 19th century to the middle of the 20th) they are very interesting and extend over the whole of Bulgaria, so we may assume that they are typical of traditional Bulgarian food. Some are connected with the Slavic people, others arrived with the proto-Bulgars and Thracians. I must emphasize that it is difficult to prove which methods are old Slavic or proto-Bulgar, and to what extent they are ancient processes originating in the Far East, Ancient Rome or Greece. In any case they exist in traditional Bulgarian food. It is not my aim here to prove where a particular method originated. This question may be answered by the future discovery of old documentary sources, and in collaboration with dialect scholars.

PRESERVATION OF FRUIT AND VEGETABLES
Among the Bulgarians the preserving of vegetable foodstuffs - fruit, vegetables and herbs - is widespread; since they can only be used in their fresh state for three or four months from the end of spring to the beginning of autumn, a large proportion of this foodstuff is preserved. The same method of preserving is used on the plain and in the mountains.

First, there is the drying in the sun of wild and garden fruits (strawberries, raspberries, bilberries, sloes and cornelian cherries, cherries, apples, pears, plums, apricots and quinces). The fruits, whole or cut in pieces, are laid in the blazing sun for a few days, then they are sprinkled with ash dissolved in water, dried again and stored. Fruits dried in this way with only water added, are used to make *ošày* (compote) which is eaten in the morning for breakfast and during Lent at other meals as well.

Water and mustard seed are poured over many of the fruits - pears, medlars, grapes - when raw. These fruits are eaten with bread and the liquid is drunk as a non-alcoholic beverage. For centuries in many mountain villages over 1,500 m above sea level - in the Rila, Rhodope and Pirin mountains - cranberries have been preserved with only water poured over them, they are remarkable for their good flavour.

In the plains of southern and northern Bulgaria and among the Bulgarians in Banat and Bessarabia, regions rich in vines, some of the fruits are piled in wood barrels and unboiled grape juice is poured over them, then they are kept in a cool place. From pressed grape juice, boiled for a long time over a low flame another preserve product is made—*petmèz*. Often it is served with small pieces of pumpkin which have been kept for a few hours in a lime (calcium) solution. Sometimes the *petmèz* is thickened by dipping a white-hot iron in it. In south-eastern Bulgaria and part of southern Bulgaria *petmèz* is made not only from grapes but also from white or black mulberries or from watermelons, and is also thickened by the insertion of a glowing iron or the addition of charcoal ash. The viscous syrup is used to sweeten wheat and maize porridge, to spread on bread and to sweeten tea and compote.

Since the early 19th century in many places in central northern Bulgaria, an area famous for its plums, plums have been baked in an oven to make what is called *pestil* (thick plum puree). From the early 20th century some of this has been exported to central Europe where it is much sought after.

For the storing of raw apples, pears, quinces, watermelons and honey-dew melons, well ripened and undamaged fruit are selected in autumn and wrapped in straw. Stored in a cool place they can be kept for up to six months. Vine branches with bunches of grapes are hung in a cool draughty place, their cut ends smeared with melted bees' wax which after hardening cuts off the air supply.

Another process is used in northern Bulgaria. Small barrels are filled with bunches of grapes (individual bunches or whole branches), unboiled grape juice is poured over them and mustard seed is added. In this way the grapes retain their flavour until spring. They are an exellent accompaniment to wine.

A widely used method for keeping vegetables - green beans (which are dried in their pods), mallows, later red paprika pods and tomatoes - is to string them up and dry them in the sun. Before being used in winter they are soaked overnight and cooked with or without the addition of other foodstuffs. The paprika pods are filled with beans, rice, chick-peas, and more rarely with pieces of meat.

Typical of the Rhodope region, and recorded since the 15th/16th centuries, is the storing of turnips, kohlrabi and Jerusalem artichokes in salt water. For almost the whole year these are used for dishes with or without meat, soups, mush and pies.

In the regions in the plain in northern and southern Bulgaria and among the Bulgarians in Asia Minor young vine leaves are preserved in early June. They are put in unglazed clay pots and salt water is poured over them. They are used for *sarmi* (leaf wrappings), which are filled with wheat grits, rice, boiled beans and, since the 18th century, pieces of meat.

One of the most typical preserved winter foods is sauerkraut. It is prepared with large barrels filled with whole cabbage heads, these are salted with sea salt, then water is poured over them and sometimes horseradish and maize grains are added. For 10 to 14 days the water is drained daily from the bottom of the barrel and poured again into the top so that the salt is evenly distributed. Then the cabbage is put in a cool place. After about a month fermentation ceases and the cabbage heads can be eaten. They are used for dishes with pork, beans, wheat, groats and rice. The whole leaves, filled with rice or meat, are used for *sarmi*.

Very widespread and characteristic are what are known as *tursii*. Green tomatoes, green or red paprika pods, carrots, and - in the southern regions - small unripe pumpkins, honey-dew melons, watermelons and cucumbers, are soaked together or separately in salt water. After fermentation they are eaten with mush, meat dishes and other meals, and often simply with bread. From the early 20th century aubergines have been preserved in some regions. They are cut up, salted, the dark, bitter juice is drained off, then they are filled with finely chopped carrots and cabbage, and equal parts of water, pure wine vinegar, salt and mustard seed are added. Here it should be noted that bottling in vinegar was very rare among the Bulgarians until the 20th century.

PRESERVATION OF ANIMAL PRODUCTS

The preserving of animal products (i.e., the processing of meat, bacon fat and fish) was also widespread until the 1950s. As many writers have observed, the main method for the long-term preserving of pork is salting. I must emphasize that in traditional Bulgarian food pork is the most important of all types of meat. Evidence has survived of the consumption of pork by the Thracians and proto-Bulgars. Traditionally in Bulgaria salting took place immediately after killing in December. The preserved meat keeps for a long time in the brine resulting from the dissolving of the salt. Meat and pieces of bacon fat are often stacked in the same vessels. Nearly all year round the fat is eaten raw or used for cooking. In some mountain regions snow is added to the preserved meat thus creating very good conditions for long-term storage.

In south-western Bulgaria (Pirin region) an interesting process for preserving pork existed until the beginning of the 20th century: salt is rubbed into legs of pork which are then wrapped in a hemp or linen cloth and laid directly in front of the front door so that anyone entering or leaving the house has to tread on the meat. After about ten days the legs are seasoned with pepper and paprika and stored in a cool place. They are considered a delicacy, keep for a long time and are eaten raw, sometimes during work in the fields in the summer.

Another process used in the same region is the cooking and storing of pieces of pork in lard.

A characteristic method of preserving meat in southern Bulgaria, with its warmer climate, is to fill a thoroughly cleaned stomach, duodenum or bladder with pounded pieces of pork salted and seasoned with wild herbs—wild thyme, peppermint, origano, savory. The finished product is hung in a dry, cool place, often on an outside wall, where it dries. It is called *stàrec, bàba* or *tàrbùf.* It keeps for a long time and is eaten raw in the fields in the summer and in the wine harvest in the autumn.

Besides this, there are other kinds of sausage made from pork which are prepared immediately after the killing. Blood sausage (i.e., the large intestine filled with pieces of heart, lungs, spleen, blood, with some coarsely ground flour, herbs and spices) does not keep so long. In some regions of central western Bulgaria lungs and liver, kidneys and heart and a part of the intestines are cut up small and cooked in lard (*kavarmà*). Sausages made from the small intestine stuffed with pieces of pork or beef, called *nàdenica* or *lukànka*, last longer. The salami *lukànka* is also very occasionally made from ox meat. I must stress that only in a few regions in southern and south-eastern Bulgaria is there evidence of preserving sausage and bacon by smoking - hanging them in the chimney hood.

In the Rhodope region another method is used: the meat is lightly salted, after two or three days the pieces are dipped in hot water, then they are hung in the chimney hood. The smoked meat is cooked with beans.

Since time immemorial in all mountain regions of Bulgaria where sheep are kept, every household kills between three and seven sheep and a few goats in autumn. The meat is cut in long pieces, salted and left for a few days in the brine that forms, then left for a few months in a dry, draughty place on the outside wall of the house. This product is known as *pastărmà* and is the favourite food of the mountain people. It is eaten raw or cooked with pulses.

In warmer areas of Bulgaria and among the Bulgarian in southern Thrace, Asia Minor and Bessarabia another method of preserving sheep meat is usual: the meat is boiled in water until it completely breaks up, the bones are removed, and the boiled up meat is put in the sheep's stomach. After it has cooled the opening is sewn up and the *sazdărmà* is coated in charcoal ash, stored in a cool place and eaten plain or cooked with eggs.

In many places on the Black Sea, the Danube and the larger rivers the preserving of fish by salting or drying is usual. The fish is sold inland by travelling fishsellers for everyday consumption and for particular annual festivals on which fish is the obligatory food.

PRESERVATION OF MILK
The methods of treating milk are also ancient and of great variety. Among the Bulgarians milk and cheese are everyday food, they are far more important and widespread than meat dishes. It is characteristic that there is a religious attitude to milk, just as there is to bread: it must not be shaken, poured away or trodden on.

The famous yoghurt made from sheep's, goat's and buffalo's milk is highly prized. According to some writers this food was introduced by the Thracians, though it is also known to the Slavs. It is prepared by adding a certain quantity of fermented milk to the luke warm milk. The oldest method of fermenting sheep's milk is by the addition of wood ants or a wild plant (called *mlečòk*), or a particular kind of thistle. The thickening and preserving of the milk also occurs through natural fermentation. In the summer when large flocks of sheep graze in mountain pastures far from the village all the milk produced in a single day is poured into large barrels and salted so that it thickens of its own accord. On the next day more milk is added which also thickens. When the barrel is full, the upper layer of fat, which stops the supply of air and thus prevents further fermentation, is poured off. This milk, called *samokiš*, is eaten in early spring of the following year.

The most widespread way of preserving milk is the production of cheese. Many writers mention that the ancient Greeks, Romans, Thracians and proto-Bulgars were familiar with cheese. In the region of Serdica (modern Sofia) fresh cheese was being sold in baskets in the 12th century. It is also a favourite food of the Bulgarians who have several methods of preparing it. The oldest of these survived until the 1930s in south-western Bulgaria. The fresh sheep's milk is poured into a large vessel in which have been placed some crushed pieces of the abomasum [the fourth stomach, which produces rennet] and a large, white-hot stone. The cheese accumulates around this stone. In some villages in the Rhodope region boiling water is used instead of a stone and stirred with a thick stick. Sometimes boiled milk and rennet are put in a rolled up, well washed and sheared sheepskin, which is then tied up. In this way the milk turns to cheese by itself. The cheese made by this method keeps very well and has a good flavour.

In the other regions of Bulgaria cheese is made by adding rennet to the milk (sheep's, cow's or buffalo's, or a mixture of them), and when it has begun to thicken the cheese is cut into pieces, the water is squeezed out and the cheese salted. Until 1920 the cheese was stored in a well-tied sheepskin or in small wooden barrels. Often curd, made by boiling whey, is added to the cheese. The cheese mixed with curd tastes very good. Curd is also eaten with pies, crumbled bread and mush, or cooked with eggs or used to fill paprika pods. Curd is made in two ways, by boiling buttermilk or cheese whey. In the Bulgarian cuisine the two sorts of curd are distinguished from each other and have different names, although they taste the same. The second sort keeps better. It is prepared in large quantities, well salted and with butter poured onto it, and stored for the winter. This curd is cooked with eggs, used to fill pies or spread on bread.

CONCLUSION

The tradition of home preserving remained on the whole relatively stable until the end of the 1950s. Up to this time the traditional popular processes had an important place in people's lives. With the gradual disappearance of traditional culture resulting from the changes in social and economic conditions and the development of industrialization the traditional preserving methods are dying out. But even today the traditional treatment of food has retained some of the old popular methods of preserving.

REFERENCES

1. Rakovski, G., *Pokazalets . . .*, vol I, Odessa, 1859.
2. Karavelov, L., *Pamjatniki Narodnogo Byta Bolgar*, Moscow, 1861.
3. Marinov, D., 'Gradivo za Veshtestvenata Kultura na Zapadna Bulgariya', in SbNU (*Sbornik za narodni oumotvoreniya i narodopis* Collection of folklore and ethnography), issue 28, vol II, Sofia 1901.
4. Sakuzov, Iv., 'Hranata na Starite Bulgari' in *Ouchitelska Misul*, year IX, 1928, issue 7, pp 429-36; 'Skotovudstvoto v Srednovekovna Bulgariya', in *Istoricheski Pregled*, 1928, issue 7-8, pp 301-28.
5. Kacarov, G., 'Zemedelieto v Drevna Trakia', in *Trakijski sbornik*, Sofia, 1933.
6. Dechov, V., 'Srednorodopsko Ovcharstvo', in SbNU, vol XIX, Sofia, 1933.

7. Zahariev, J., 'Kyustendilsko kraishte', in SbNU, vol XXXII, Sofia, 1918.

8. Titorov, J., *Bulgarite v Besarabiya*, Sofia, 1935.

9. Slatarov, A., *Prigotvjane na Razlichni Vidove Sirene v Bulgariya*, Sofia, 1921.

10. Vakarelski, H., *Etnografiya na Bulgariya*, 1974, 'Bit na Trakijskite i Maloazijski Bulgari', Sofia, 1935.

11. Telbizov, K., and Vekova-Telbizova, M., 'Traditsionen Bit i Koultoura na Banatskite Bulgari', in SbNU, vol LI, Sofia, 1963.

12. Vakarelski, H., 'Veshtestvena Koultoura na Strandzhanskata Oblast', in *Sbornik Strandzhanska Ekspeditsiya*, Sofia, 1957.

13. Primovski, A., 'Bit i Koultoura na Rodopskite Bulgari', in SbNU, vol 54, Sofia, 1973.

14. Krusteva, G., 'Narodna Hrana i Hranene', in *Sbornik Dobrudzha*, Sofia, 1974, pp 249-61.

15. Radeva, L., 'Hrana i Hranene', in *Sbornik Pirinski Kraj*, Sofia, 1980, pp 347-68; 'Hrana i Hranene', in *Sbornik Plovdivski Kraj*, Sofia, 1986, pp 166-88.

16. Radeva, L., chapter 'Hrana i Hranene', in *Etnografia na Bulgariya*, vol II, Sofia, 1983, pp 288-300.

17. Radeva, L., 'Nahrung und Ernärung' from the districts of Blagoevgrad, Plovdiv, Pazardzhik, Smolyan, Haskovo, etc, in *Arhiv na Etnografskiya Institout i Mouzej pri BAN* [BAN = Bulgarian Academy of Sciences].

INSIGHTS INTO THE PROBLEM OF PRESERVATION BY FERMENTATION IN 6TH CENTURY CHINA

Françoise Sabban

From a study of a 6th century agricultural text, the Qimin Yaoshu, *it is clear that for the Chinese, fermentation was a vital process. The rules and precision employed in its execution were indicative of its status.*

It is well-known that foods preserved by fermentation are important in China today. Almost everywhere in that country one can sample local 'wines' made by the fermentation of glutinous rice, sorgho or millet and many families still make their own *paocai* or *yancai*—Chinese versions of what we know as sauerkraut.

The basic and most common condiment used throughout China is soya sauce, itself a product that results from the fermentation of soya beans and wheat. It is difficult to imagine Chinese cooking without soya sauce and yet the record shows it is a relatively modern adjunct to the list of Chinese condiments.

In order to understand the role and importance of fermented foods in the past I have undertaken a study of a 6th century agricultural text, the *Qimin Yaoshu* (from now on, QMYS). I chose the QMYS because it is one of the oldest surviving texts on agriculture[1] and cooking. Recourse to an agricultural text may at first appear strange when speaking of cuisine, but one fourth of the QMYS is dedicated to food preparation; indeed, this text had so great an influence on later culinary texts that one could argue that it is the first great Chinese culinary treatise.[2]

The QMYS[3] was written between 533 and 544 by Jia Sixie, about whom we know nothing except that he was a middle-ranking official with agricultural experience in the climatic conditions of northern China. The title of his treatise has been variously translated into English as 'Essential Ways for Living of the Common People'[4] or 'Essential Techniques for the Peasantry'.[5] The author introduces his monumental work in these terms:

> I have gleaned material from traditional texts and from folksongs. I have enquired for information from old men and learned myself from practical experience. From ploughing to pickles there is no domestic or farming activity that I have not described exhaustively. I call my book 'Essential Techniques for the Peasantry'. In all, it comprises 92 chapters divided into ten books.[6]

Among these ten books, the whole of books VIII and IX and the major part of book VII concern the domestic economy of food; in other words, 25 of the 92 chapters are dedicated to food preparation. There are recipes both for dishes to be consumed immediately or soon after their making, and for others that are meant to be kept a certain time before being eaten. Hence, portions of the text resemble a cookbook while others are closer to treatises on domestic economy.

Of the 25 chapters in question, 12 are devoted to foods preserved by fermentation. Eight deal with the preparation for 'starters', alcoholic beverages and fermented condiments such as relishes and vinegars. The rest is devoted to other preserving methods for meats, fish and vegetables. In addition, there are some scattered recipes for food preservation in chapters treating the cultivation of certain fruits or vegetables; and, in book VI which deals with husbandry, there is a good description of milking practices in the course of which a method of fermenting milk to make a yoghurt-like product is described.[7]

FERMENTED PRODUCTS: DRINKS AND FOOD
STARTERS. The fermentation process is not always a simple linear operation. Many times, and especially when alcoholic beverages are concerned, the operation requires a 'starter' which itself is the result of a previous fermentation. The various starters obtained through fermentation of cereals are then the requisite elements which permit further food fermentation. Unlike the 'starters' used in making bread, which can be taken from a previous batch of dough, Chinese starters have to be made from scratch every time and involve a chain of complex operations similar to those employed in preparing other fermented foods. The starters for alcoholic beverages are called qu[8] and the two others, used for relishes and vinegars, huangyi (yellow mould) and huangzheng[9] (yellow steam).
ALCOHOLIC BEVERAGES. In the QMYS, alcoholic beverages are called jiu[10] and most of them are obtained by the fermentation of a cereal, (glutinous and non-glutinous rices and millets) induced by a qu starter.[11] We will later examine the special terminology which Jia Sixie uses to describe 'brewing; in his descriptions every stage is particularly well explained.
RELISHES. There are two families of relishes described in the QYMS: the chi and the jiang. The jiang[12] seem to be the most diversified. Such relishes could be of vegetal or animal origin (soya beans, wheat, elm-pods, meat, fish, shrimps, crabs were all used) and fermentation was induced by the 'yellow' starters huangyi and huangzheng. The chi condiments were made quite exclusively by the natural fermentation of soya beans.[13]
VINEGARS. Vinegars, called cu,[14] are the result of an acetic fermentation with the yellow starter huangyi. They can be made of various basic ingredients which include cooked cereals like millet, rice or barley, wheat flour, wine dregs, spoiled wine, sour wine, honey and a kind of Chinese apricot.
FERMENTED STARCH PRESERVES ZHA. Zha preserves were principally made from fish (whole, cut into pieces or minced) that was fermented with cooked rice and flavoured with different aromatic plants and spices.[15]
THE FU PRESERVE. These preserves could be made of virtually any meat or fish. Meat is generally put in a kind of aromatic brine before being hung under the rafters to dry completely. Made in winter, it would have to keep for several months until summertime.[16]
THE LACTO-FERMENTED VEGETABLES ZU[17] Many vegetables' preparations called zu could be considered as fermented food quite similar to

what is called 'sauerkraut' in Northern Europe.[18] Vegetables like Chinese cabbage, mallow, kohlrabi, mustard greens, etc, are put in preserving pots with salt, brine or rice mush where fermentation occurs with or without the addition of a 'yellow' starter. *Zu* recipes for keeping fresh vegetables make up a full chapter of Book IX. The difference between preserved and fresh vegetables is not always clear since, in addition to designating the preserved vegetables described above, the term *zu* is also applied to vegetables simply parboiled and prepared as salad. In the QMYS *zu* seems to be designating a state somewhere between raw (*sheng*) and cooked [=boiled] (*ru*). But even if *zu* salads are lightly cooked they are considered fresher, 'rawer' than the *ru* preparations which have to be thoroughly cooked [boiled].[19]

THE BASIC STAGES OF THE FERMENTATION PROCESS
In the QYMS the various stages of fermentation are described according to a unique system in which all features do not necessarily appear together for each operation:
   Choice of the right time
   Choice of the right place
   Care in the selection of the raw materials
   Quantities in due proportion
   Rules of hygiene
   Magical rules
   Time of 'ripening'
   Time of preservation
   Advice (uses, preservation, etc)
Every step is detailed with striking precision. The quantities employed, the best place for each operation to be carried out, even the smallest gestures are minutely described. The author sometimes even explains why he recommends certain actions, telling his reader what problems might arise if he does not follow his advice. This precision is made possible through the use of a system of 'touch stones' based on sense perception: taste evaluation is important as is colour perception; touch intervenes when judging consistencies, and temperatures are measured with the hand. The temperature that always serves as reference is that of the human body (precisely, that of the armpit). These measurements permit step by step control of the fermentation process.

Another remarkable characteristic of these descriptions is their obsession with hygiene. The place in which fermentation occurs must be clean, and so should be the men who carry out the fermentation (they are asked to brush their finger nails, for example). Naturally, cooking and fermenting utensils have to be perfectly clean or else new. It is clear that the author knows that the success of the whole operation depends to a large extent on respecting these hygienic customs.

THE LEXICAL EXPRESSION OF FERMENTATION

Shi Shenghan, editor and translator in modern Chinese of the QMYS, has drawn attention to the advanced technology employed by Jia Sixie to induce fermentation and, generally speaking, the large scope of his knowledge and technical skill. One proof of Jia Sixie's technical expertise is the specialized vocabulary he uses. As mentioned previously, Jia Sixie's descriptions, on a lexical level, are highly precise and clear - with the exception of those portions of the text corrupted by successive copies and editions.[20] It is not clear when and how the kind of technical vocabulary used by Jia Sixie came into existence. According to some scholars, already in very early texts like the *Ritual of Zhou* (4th century BC), the various fermentation stages of wine are described with what seems to be a technical vocabulary.[21]

What is remarkable is that Jia Sixie uses a technical vocabulary specific to the context of fermentation in conjunction with a general one whose items take on a technical meaning, different from their customary one, in the context of his discussion of fermentation.

In the QMYS specific products and actions are referred to by words from the technical terminology of fermentation. This terminology concerns either predicative terms or substantive terms. Substantive terms and the products they referred to were already in existence when the QMYS was being written whereas the vocabulary of actions and operations (i.e. predicative expressions) seems to have been less extensive and was probably less fixed.

We have seen that every fermented food or beverage has its own appellation which in itself defines a precise fermentation process. The same is true of the names of the various starters. In addition to these, other by-products and microorganisms, which can be useful or dangerous during the fermentation process depending on the circumstances, are precisely identified by name. For example, when a specific mould is expected to appear (most often a necessary one) it is designated by the generic term *yi* (literally 'coat, dress'). On the other hand the undesirable white mould that can appear on the surface of vinegar[22] when the fermentation goes awry is referred to as *baipu* (white *pu*).[23] One of the by-products, the dregs of either wines or vinegars is called *zao*.[24] In addition to the vocabulary referring to these products, whether expected or unexpected, useful or useless, there exists an impressive set of predicative expressions describing the different stages of the various types of fermentation. However, as previously stated, there are only two verbs in the technical terminology specific to fermentation: *niang* which means 'brew or prepare an alcoholic beverage'[25] and *tou* which refers to the action of successively adding grains of a cereal to the soaked and activated starter until the alcohol produced in the liquid has reached such a level that the fermenting microorganisms can no longer survive; the brew is then ripe. With these two exceptions, the verbs employed are ordinary verbs which take on a special meaning in this particular context. For example, to qualify the fermenting power of a starter in wine making, it is said that one dose of it can 'kill' (*sha*) x amount of grain. The verb *xiao* which usually means 'disappear', here describes the digestive action of the same starter on the cereal. As for *wo*, which normally means 'to sleep, or to go to bed' it is used in relation to the fermentation of relishes and vinegars to signify 'putting something in a safe place at a constant temperature'. The idea of 'keeping' a product for a certain time will be expressed by the term *ting* which, in ordinary language, means 'to stop, to pause'. The bubbling of a liquid in fermentation will be described as 'boiling' with the verb *fei*, generally used for the boiling of water. When the whole operation is finished and the fermented food is ready to be consumed, it is generally said that it is *shu*. In modern Chinese texts, this adjective can be translated by 'ripe' or 'cooked'. In several ancient technical texts, the termination of an operation, the transmutation of a product from its original to its finished state is expressed by this term. So it is for the accomplished fermentation process.

WINE-BREWING: THE MODEL
If we consider the importance accorded to their preparation and also the fermentation terminology, it is obvious that, for Jia Sixie, the true and quite sacred fermentation process of greatest importance was that associated with the brewing of alcoholic beverages. The third of the 12 chapters dealing with fermented food preserves is exclusively dedicated to the making of starters (*qu*) and wines. And it is noteworthy that the first recipes for fermented foods listed in Book VII are for alcoholic preparations. The brewing of wines was without any doubt very important in the domestic organization of a big estate like Jia Sixie's and also, as we know, in social and religious life at that time.[26] We note that, in requesting help of gods for the preparation of only one *qu* starter indispensable in making wine,[27] Jia Sixie reveals to us the importance he himself attached to alcoholic fermentation.

As for terminology, we have stated that the richest set concerns wine-brewing and that only for wine-brewing does there exist a specific term meaning 'to brew'. Furthermore when there is a comparison between two ways of fermenting, the term of reference is always that used for the brewing of alcoholic beverages, as if it was the actual model to be followed in other cases. We have a wonderful lexical example of this tendency in a recipe of sauerkraut where Jia Sixie uses a special Chinese character transferred from the character *niang*, normally applied to wine-brewing, to which is added the 'herbal' key, conferring upon the resulting character the specific and unique meaning of 'to induce a fermentation process with vegetal ingredients'.[28]

CONCLUSION
Any conclusion about the 6th-century Chinese conception of fermentation must take into account the hierarchy between the various food preparations described in Jia Sixie's QMYS. This hierarchy emerges from a close examination of the text: What is treated first? The starters. What are their

uses? They are used to induce alcoholic fermentation in cereal mushes producing wines which we know were important at that time. For Jia Sixie the 'real' or 'true' fermentation, that which required the greatest care and even intervention by gods—whose accomplishment, therefore, did not depend entirely on human factors—was wine-brewing.

Moreover, as the vocabulary shows, the fermentation process is a kind of slow ripening and may be compared to the development of a child in the belly of his mother.

Paradoxically, however, pregnant women are often considered dangerous for the fermentation process. If the presence of such a woman has spoiled a relish or a vinegar, our treatise includes recipes for neutralizing the damage: spoiled vinegar can be restored to its normal state by putting into it earth taken from the grooves cartwheels leave in the roads! Just as one is advised never to use a pot previously employed in a fermentation process, a pregnant woman should not be allowed to interfere with it since she too is in fermentation, and her fermentation could hinder the fermentative phenomenon specific to the production of these special types of food and drink.

ACKNOWLEDGEMENT

I would like to thank Mary and Philip Hyman for their kind help in writing this paper in English.

NOTES

1. This is not the first Chinese treatise on agriculture: we know of the existence of the *Fan Shengzhi shu*, an agricultural treatise from the Han dynasty (206 BC - 221 AD) and the famous calendar *Simin yueling* by Cui Shi (about 150 AD), which have both disappeared but which are cited on occasion in later works and especially in the QMYS. See Shi Shenghan (ed), *On Fan Sheng-chih shu. An Agriculturist Book of China Written by Fan Sheng-chih in the First Century BC*, Science Press, Peking, 1963, 68 pp; and Shi Shenghan (ed), *Liang Han nongshu xuandu, Zhongguo nongxue puji congshu* Nongye chubanshe, Peking, lst edn 1962, 1979, 50 pp. The *Qimin yaoshu*, however, is the most ancient of the agricultural encyclopaedias that has been preserved in its entirety.

2. I do not want to suggest that it is impossible to find Chinese texts containing information on food habits prior to the 6th century: on the contrary there are many, but the QMYS is the first text giving purely technical instructions and recipes. Later, from the Tang dynasty (618-907) onwards, culinary texts were dissociated from agricultural matter, hence the QMYS was both the first and the last text dealing with agriculture and cooking. For further information on this point see Yukio Kumashiro, *Recent Developments in Scholarship on the CH'IMIN YAOSHU in Japan and China*, The Developing Economies, Tokyo,

Developing Economies, Tokyo, 1971, IX/1-4, pp 422-48. On cooking in early Chinese texts see David Knechtges, *A Literary Feast: Food in Early Chinese Literature*, Journal of American Oriental Society, 1986, 106.1, pp 49-63; and also of course K.C. Chang, 'Ancient China', and Ying-shih Yu, 'Han', in K.C. Chang (ed), *Food in Chinese Culture, Anthropological and Historical Perspectives*, Yale University Press, 1979, pp 23-85.

3. In the course of preparing this paper the following editions were consulted:
    Miao Qiyu (ed), *Qimin yaoshu jiaoshi*, Nongye chubanshe, Peking, 1982, 870 pp.
    Shi Shenghan (ed), *Qimin yaoshu xuandu ben*, Nongye chubanshe, Peking, 1st ed, 1961, 1981, 636 pp.
    Shi Shenghan (ed), *Qimin yaoshu jinshi*, Xibei nongxueyuan gu nongxue yanjiushi congshu, kexue chubanshe Peking, 1958, 721 pp.

4. Shi Shenghan, *A Preliminary Survey of the Book ch'i min yao shu, An Agricultural Encylopaedia of the 6th Century*, Science Press, Peking, 1962, 2nd edn, 107 pp.

5. Francesca Bray, Agriculture, Part II, in Joseph Needham, *Science and Civilisation in China*, Vol 6, Biology and Biological Technology, Cambridge University Press, 1984, pp 55-9.

6. Francesca Bray, *op cit*, p 56.

7. See Francoise Sabban, *Un savoir-faire oubli: le travail du lait en Chine ancienne*, Memoirs of the Research Institute for Humanistic Studies, Kyoto University, Zinbun, 1986, 21, pp 31-67.

8. Miao Qiyu (ed), *Qimin yaoshu jiaoshi*, op cit, book VII, chapters 64, 65, 67. QMYS contains 9 different recipes of *qu* starters. They are obtained by the fermentation of wheat 'cakes' (raw, boiled or toasted wheat grains associated or not) which sometimes include magical prescriptions. See in particular the passage on the preparation of the 'heavenly *qu* starter' in chapter 64, p 358.

9. Miao Qiyu (ed), *Qimin yaoshu jiaoshi*, op cit, book VIII, chapter 68, *huangyi* and *huangzheng*, respectively 'yellow mould' and 'yellow steam' can be called the 'yellow' starters. In comparison with the *qu* starters their fermentation process is rather simple. In both cases, wheat is the raw material used. For *huangyi*, whole grains are soaked in water until they become sour; they are then steamed and put on a mat until they start to mould. For *huangzheng*, wheat flour is first kneaded with water and then prepared in the same way as *huangyi*.

10. In the QMYS the alcoholic drinks are all designated by the same term *jiu*. From now on, we will use the word 'wine' to translate this morphem *jiu* which in modern Chinese still functions like a generic term designating any kind of alcoholic beverage, distilled or not, such as grape wine, brandy, or beer. In 6th century China, *jiu* obviously designates alcoholic beverages obtained through grain fermentation. It is generally admitted that distillation in China occurred under the Yuan dynasty (1206-1368), but there is now a debate on this

topic, since some specialists argue that it was probably practised at a much earlier date. For further information se Xing Runchuan, *Gudai niangjiu jishu yu kaogu faxian* [Archeological Discoveries and Techniques of Brewing in Chinese Antiquity], Kejishi wenji, 1984, 9, pp 93-98.

11. Miao Qiyu, *op cit*, book VII, chapters 64, 65, 66, 67, pp 358-407. Out of a total of 40 recipes for alcoholic beverages in the QMYS, only three are not instructions for fermenting grains: one describes how to make a fruit wine and the other two a medicinal wine.

12. Miao Qiyu (ed), *op cit*, book VIII, chapter, 70. The QMYS contains 14 recipes of *jiang*. The history of the relishes referred to as *jiang* is interesting because, though the name still exists today, the reference has changed and in a sense it represents the evolution of the basic flavours of Chinese cuisine. Today *jiang's* meaning is 'very rich', but it is primarily used as the general appellation for fermented relishes made of soya beans and wheat. The morphem *jiang* is also part of the compound word *jiangyou* [*jiang*-oil] 'soya sauce' and it can enter in the designation of some preserves which include soya sauce in their composition. Furthermore, by extension *jiang* designates various kinds of pureed foods like tomato-paste (*fanqie jiang*), jam (*guozi jiang*) or mustard (*jiemo jiang*). The most ancient mention of this term appears in the *Ritual of the Zhou* (4th century BC) (see Hong Guangzhou, *Zhongguo shipin keji shigao* [Historical Survey of Chinese Food Technology]. Zhongguo shangye chubanshe, Peking, 1984, p 90) where it seems to refer both to a kind of vinegar and to mincemeat, a meaning very different from that of today. In many texts prior to the QMYS we find the mention of *roujiang* [meat-*jiang*], both the word and the product seem to have disappeared by the 6th century since they are not mentioned in QMYS. It is only in a text dating from the the 1st century BC (*Jijiupian*, see Hong Guangzhu, *op cit*, p 93) that *jiang* acquires the new meaning of relish made of fermented soya. This discussion of the evolution of a Chinese cookery term is more to the point than may at first appear. It throws light on the origin of soya sauce which, contrary to what is normally assumed, was not the result of a deliberate preparation but only the by-product of making certain relishes. Before the Song Dynasty (960-1279) soya sauce was merely an 'oil' that oozed out in the course of making a *jiang* relish; indeed, it was called either *jiangqing* [*jiang*-clear] 'the clear part of *jiang*' or *jiangzhi* [*jiang*-liquid] 'the liquid part of *jiang*' and the fact that it had no name of its own, so to speak, indicates its secondary importance at the time.

13. Miao Qiyu (ed), *op cit*, book VIII, chapter 72, pp 441-448. There are 3 recipes of *chi* in the QMYS, two made of soya beans and one made of wheat. Nowadays a condiment called *chi* still exists. It is also the result of a natural fermentation of soya beans but, unlike the product described in the QMYS where the last stage of the process is a fermentation in rice chaff, today it is made in brine.

14. Miao Qiyu, *op cit*, book VIII, chapter 71, pp 429-441. For the history of vinegar in China and the evolution of the word*cu* (now written with another Chinese character), see Hong Guangzhu, *op cit*, pp 112-128. In QMYS, we find more than 20 recipes for making vinegars. In a few of them, clearly mentioned as coming from prior texts, vinegars are not called *cu* but *kujiu* [bitter-wine] - or should we say spoiled wine?

15. Miao Qiyu (ed), *op cit*, book VIII, chapter 71, pp 429-441. 7 recipes in QMYS are classified as *zha*. 6 are made with fish and one with pork. In fact, as Shi Shenghan (*Preliminary Survey ... op cit*, p 88) suggests, *zha* 'was originally started in the Yangtze valley and fishes alone were thus conserved' - the Chinese character transcribing*zha* tends to support this since it indeed contains the 'fish' key.

16. Miao Qiyu (ed), *op cit*, book VIII, chapter 75, pp 459-463. We do not know exactly which kind of fermentation is involved in the 7 recipes of*fu* preserves; maybe they can be compared to the European cured meat preparations. It was recommended that they be made between the last month and the first two months of the year, the best time being the last month of the lunar year called in Chinese *la(yue)* [*la*—month]. For this reason, the title for the*fu* preserves chapter in the treatise is*fula*. Today this kind of meat preserve is still called *larou* [*la*—meat].

17. The term vegetable also designates some fruits such as melons or pears.

18. Miao Qiyu, *op cit*, book IX, chapter 88, pp 531-46. Of the 41 recipes in the *zu* chapter about 16 could be strictly classed as 'sauerkraut', in other words 39%.

19. See for example the first recipe of chapter 78 (Miao Qiyu,*op cit*, p 531): '*Zu* is still green, wash it to put off salt, boil it as *ru*, it makes no difference with [cooked] fresh (raw) vegetable.'

20. For further explanations on the history of this text see the excellent appendix of Miao Qiyu's book (pp 733-870), dedicated to the various problems posed by the successive reproductions and editions of an ancient text.

21. See Hong Guangzhu,*op cit*, pp 126-156.

22. See for example the recipe for 'Barley vinegar' in Miao Qiyu (ed),*op cit*, book VIII, chapter 71, p 431.

23. The fundamental difference between the *yi* and the *baipu* moulds is expressed in the terms themselves. Whereas the word *baipu* can be applied only to vinegar fermentation and is limited to the sole context of fermentation, *yi* can refer to any kind of mould and has a complex semantic content. Its first meaning (coat, dress), referring to things of the ordinary world, can take on specialized meanings that must be deduced in special contexts like fermentation; hence, the first, general meaning always stays behind the derived one.

24. The great majority of wines and some vinegars are pressed to eliminate their solid part called *zao*, which is sometimes used in other preserved food recipes,

like one for a *zu* preparation of melon, because of its taste and fermentative powers (see Miao Qiyu (ed), *op cit*, book IX, chapter 88, p 533).

25. It is also used to qualify the fermentation of a vinegar. When combined with the verb *xia* (literally 'go down' with brewing), the compound *xianiang* means 'start fermentation'. Today, *niang* still bears this acception with the derived meaning of 'to foment secretly'. It also refers to the work of bees producing honey.

26. 'Since the earliest times, drinking *chiu* ('wine') was a basic part of Chinese living, whether in the highest rituals, where aromatic beverages were proffered on state altars to the gods, or in any casual moment of relaxation from work or worry by the lowliest member of society' (Edward K. Schafer, 'Tang' in K.C. Chang (ed), *Food in Chinese Culture*, *op cit*, p 119). There is an ancient and long tradition in Chinese literature of 'drinking-poetry'. Many of these poems praise the wonderful intoxicating properties of wines. That suggests the importance for a wine to give ebriety and explains some notations in our treatise on the ability of certain wines to give a deep drunkenness. According to Wan Guoguang, *Jiuhua* (*On Wine*, Kexue puji chubanshe, Peking, 1987, p 101) under the Northern Wei (4th-6th centuries AD) the literati particularly appreciated wine drinking for medical reasons but also for political reasons. Sometimes, drunkenness was the best way to escape troubles.

27. Book VII, chapter 64 (Miao Qiyu, (ed), *op cit*, p 358) begins with the recipe of the 'Three wheats *qu*' which is the most detailed and the longest recipe for *qu*. Besides the description of the technical process there is the painting of a complex ritual with a supplication to the gods for the success of the whole operation, said in a loud voice by one of the responsible members of the family. No other recipe contains such a strange practice.

28. See Miao Qiyu (ed), *op cit*, p 532 and note 5, p 542. We cannot tell whether or not Jia Sixie coined this character, but it would be interesting to determine when exactly it was first used. Miao Qiyu cites a dictionary dating from the 3rd century AD which already contains it.

APPENDIX: INDEX OF CHINESE CHARACTERS

baipu 白醭

chi 豉

cu 酢

fanqie jiang 番茄醬

fei 沸

fu 脯

fula 脯臘

guozi jiang 果子醬

huangyi 黃衣

huangzheng 黃蒸

Jia Sixie 賈思勰

jiang 醬

jiangqing 醬清

jiangyou 醬油

jiangzhi 醬汁

jiemo jiang 芥末醬

jiu 酒

kujiu 苦酒

larou 臘肉

la(yue) 臘月

niang 釀

paocai 泡菜

Qimin yaoshu 齊民要術

qu 麴

roujiang 肉醬

ru 茹

sha 殺

sheng 生

shu 熟

ting 亭

tou 酘

wo 臥

xianiang 下釀

xiao 消

yancai 腌菜

yi 衣

zao 糟

zha 鮓

zu 菹

THE ROLE OF PRESERVED FOOD IN A NUMBER OF MEDIEVAL HOUSEHOLDS IN THE NETHERLANDS

Johanna Maria van Winter

A description is given of the methods of preservation known in medieval times in the Netherlands, and of how and when they were most effectively employed.

To be able to determine the role of food preservation in medieval times in the Netherlands, one must find out which methods of preservation were practised and to which foods they were applied. Only when this has become clear can one compare the relative importance of preserved food with that of fresh food in the same households.

Our sources are too few, however, to arrive at any exact figures or percentages. Most of our information comes from kitchen accounts and there are almost none of those left for the period preceding the 14th century. Excepting the oldest city accounts, those of Dordrecht, which date from 1284/5 and 1285/6 and which provide some evidence, there are only accounts dating from the 14th and 15th centuries for the period of the Middle Ages. The households concerned are those of a few secular and ecclesiastical princes, such as the Count of Holland and the Bishop of Utrecht; that of the customs officer at Lobith, who collected toll on the Rhine on behalf of the Duke of Gueldres; and, finally, those of various monasteries and other religious institutions. As for the latter, I have consulted - for the purpose of this inquiry - only the accounts of the Abbey of Egmond and of the House of the Teutonic Knights in the city of Utrecht.[1] Household accounts of burghers in the medieval Dutch towns and cities have, as far as I know, not been discovered. Neither have any accounts of the landed nobility been found, excepting those of the lords of Blois, who derived from a 14th century collateral branch of the Holland-Hainault countly dynasty, and who possessed properties in the neighbourhood of the towns of Gouda, Oudewater and Schoonhoven.[2]

Another problem one comes up against in this kind of inquiry is the fact that the sources, apart from being scarce, were not designed to show the relative importance of fresh and preserved food. The clerk or chief steward, who drew up such accounts, was primarily interested in the cost of what was bought and stored, not in the question of whether fresh or preserved food was to be used in the kitchen. Only if the preservation process itself entailed costs was it mentioned: the purchase of salt for the slaughter, for instance, was noted down. Whether meat bought elsewhere was fresh or salted is, however, not specified. All we can say is that certain terms seem to point to preserved food: for instance, 'meat bought per barrel' or 'a portable barrel of beef'.[3] Examples of fish are: 'a basket of dried plaice', 'basket-herring' (i.e. dried, lightly salted herring), 'red herring' (i.e. smoked herring) and 'stockfish'.[4] But it remains dubious whether the omission of such a specification regarding meat and fish justifies the assumption that it was therefore fresh.

The diet of a well-to-do Dutch household in the late Middle Ages consisted primarily of meat, usually beef, from oxen and cows. These were partly raised in the Netherlands and partly imported from Denmark and Schleswig-Holstein and then fattened up in the Netherlands. Some well-to-do households had their own pastures and cowherds to keep the animals alive until they were selected for the kitchen. This was the case, for instance, in the household of the Count of Holland, in that of the Teutonic Knights and, to a certain extent, in that of the toll-house at Lobith.[5] The purchase of hay and other fodder for these cattle, as well as the cost of herding, was registered. We may assume that after the slaughter, the meat was consumed within a reasonably short time, without having been salted first. Slaughter, in fact, took place throughout the year, not only in October and November. The skins were sold and tallow candles (*ruetenkaarsen*) were made from the abdominal fat.[6]

There were households in which beef was bought every week, for instance that of the Bishop of Utrecht in 1377/8, for the duration of his residence in the city.[7] In addition to the provision of fresh meat, there was sometimes an annual slaughter for storage purposes, as at the toll-house of Lobith.[8] In this case, the meat must have been salted; in the former case, at the bishop's household, it is likely that fresh meat was bought every week at a butcher's shop in the city of Utrecht. When the bishop was staying at one of his castles outside the city, live cattle were sometimes driven there. The account of 1377/8 mentions the fact that seven cows on their way to the castle Stoutenburg had somehow become stuck near Amersfoort and had to be slaughtered; their skins were sold.[9]

Not only beef but also mutton was eaten in the Netherlands, occasionally even in large quantities. Sometimes a large number of sheep were bought at once for the kitchen and sometimes a few were bought per week, as for the Bishop of Utrecht.[10] Sheep, however, have only about one-tenth as much meat as cows, and therefore they probably did not constitute the bulk of the meat diet in these regions. It is not clear from the accounts whether these sheep were also pastured on household grounds before being slaughtered; in at least one case, that of the steward of the Count of Holland, it appears that the skins were sold afterwards.[11] In this household the mutton was probably consumed in its fresh state. Ostensibly, this was also the case in the toll-house of Lobith, even though the skins were not registered in its accounts. Whether this state of affairs corresponds to that in other households in this period must remain uncertain.

The third sort of quadruped which was bred for regular consumption was the pig. I have the impression that its meat was preserved and that slaughter did not take place immediately before consumption, except for the few pigs that were roasted fresh, as at the toll-house of Lobith.[12] Pork chops or steaks do not appear in the accounts, but sides of bacon, joints and hams are mentioned.[13] In the cases that boars (male pigs) were bought together with a large quantity of salt—as in the house of the Teutonic Knights in 1377/9[14]—one can assume that the meat was salted. Whether bacon and ham were sometimes smoked, I cannot say. As for the making of pork sausages, there is very little evidence in the accounts that are extant.[15]

Cookery-books or collections of recipes might have provided information about this, but none have yet been discovered in the Netherlands for the 14th and the larger part of the 15th century. The first evidence is a manuscript of 1480 and, thereafter, a printed cookery-book of about 1510.[16] In the manuscript, there is no mention of ordinary pork sausages, but only a recipe for a mixture of suckling pig's meat with hard-boiled eggs and spices, which could be used as a stuffing for a white sausage and other stuffed dishes.[17] In the printed cookery-book too, pork sausages are absent. Of adult pigs, the sides of bacon and the lard were especially used, but it is not clear whether these were salted or smoked.

Besides the domesticated quadrupeds, bred for human consumption, there was also, of course, game in the form of wild boar, deer and roes. These appeared now and then on princely tables as gifts or trophies of the chase.[18] I assume that this meat was not preserved, but eaten fresh, that is to say, after it had had time to cook and harden.

Poultry was another category of meat. Of this, especially chickens were eaten on a large scale - fresh, I assume. The same is true of capons, ducks and geese, which were also consumed regularly.[19] On the tables of the well-to-do, herons and bitterns, caught by huntsmen, also appeared fairly often;[20] but again there is no indication that these were preserved.

Besides meat, medieval people ate a great deal of fish—if they could afford it, for fish was not exactly cheap. The church prescribed long periods of fasting in which one was not supposed to eat any meat, animal fat, eggs or dairy produce. Also, one was supposed to refrain from eating meat for two days a week throughout the year; on these days dairy produce and eggs were, however, allowed. Instead of meat, the people in the Netherlands ate fish, for it was—and still is—a river delta country, in which sea fish and fresh-water fish were available in the markets. Fish, however, spoils even faster than meat, so that, besides fresh fish, there was a demand for preserved and dried fish.

There were, for instance, baskets of dried (and lightly salted) herring to be had, as well as smoked herring (i.e. red herring) and herring in barrels, which were strongly salted. In the course of the 15th century, more cured herring was served, called *kaakharing*, this being herring which had been gutted and placed in salt on board the fishing ship. Earlier, the fishermen had brought their catch to shore before salting it; now, doing this on board, it was possible to fish further away from the coast and to stay away longer.[21] The accounts do not make clear which methods were used, but they do show that there were various kinds of herring; smoked, salted and 'green', or almost fresh, herring among others.[22]

Another kind of fish which was put on sale, fresh as well as preserved, was cod. As stockfish, cut open and dried on sticks, it was imported from

Bergen in Norway and eaten in the Netherlands—sometimes almost on a weekly basis, as in the toll-house at Lobith.[23] However, cod could also be caught off the Dutch coast and salted, as was done at the order of the Count of Holland in preparation for a military campaign against the Frisians in the summer of 1345. The steward made a detailed account of this provisioning and he was right, for the enormous quantity (7342 fish) was intended to feed a whole army. To preserve it all, 20 *mud* of salt were necessary.[24]

Red herring, too, was laid in for that campaign on a large scale: 10 *last* and again 45 *last*, a *last* being about 11000 fish. Somewhat less, but still considerable in quantity, were the 45 barrels of preserved eel, largely salted, for a smaller part smoked.[25] Somewhat earlier, even larger supplies of fish and also meat had been bought in preparation for a siege of the city of Utrecht, which took place in July 1345.[26] The count survived this siege, which he lifted on 22 July having been able to capture the city, but he fell in the battle against the Frisians on 26 September 1345.

Outside war-time, it may be assumed, people ate fresh cod and eel, depending upon where they lived: along the sea coast it was cod, and along the rivers and brooks, eel. In 1344/5, a monastery such as that of Egmond, however, bought 15 *tael* eel (about 3000 fish), which must have been prepared for conservation. At the toll-house of Lobith, half a cask of eel was bought in May 1428.[27]

What else did people eat, besides meat and fish? In the first place, bread, made of rye or barley, and, for special occasions, of wheat. Bread-making however, cannot be counted as a conservation process. Besides bread, there were pulses, especially peas.[28] These were preserved by drying, just as were subtropical fruits: figs, dates, raisins and currants. Especially in the long fasting period before Easter, the latter fruits, together with almonds and other nuts, were consumed in well-to-do circles.[29] Of our modern practice of preserving fruit in sugar, there is no trace, neither in the medieval accounts nor in the cookery-books. Plums, cherries and berries were apparently not preserved, even though sugar had been available since the 13th century at least. Seeing that archaeologists have found stones and pips of these fruits in medieval cesspits,[30] I wonder what was done with these fruits: did people eat them raw, or were they served stewed in pies and compotes?

Since, in the medieval period, raw fruit was believed to be bad for one's health, I also wonder what was done with apples and pears. There are recipes from the 16th century in which apples and pears are used: stuffed apples, apple sauce, apple pie, apple turnovers and stewed pears in a spiced wine-sauce.[31] In the autumn of 1377, the Bishop of Utrecht had pears, to eat fresh, brough to his private room every day, and in the spring of 1378, a few times, apples.[32] I have not found such a practice in other households, however.

As far as I can see, apples in large quantities were probably used in another way: they were made into *verjus*, a kind of vinegar. In the House of Teutonic Knights in Utrecht, apples were bought for the cellar several times in 1377 and 1378.[33] Since this cellar was used especially for fluids, I conjecture that these apples were made into *verjus*. Apple juice and apple cider are also possibilities, but these fit less well into the Dutch dietary pattern, in which beer or ale was the drink of the common people and wine was imported for the well-to-do, together with the more expensive German hop-beer.[34]

Pickling in vinegar must have been a well-known method of preservation, but we find little direct evidence of this use: only that a sturgeon, which had been given as a present to the Lord of Blois, was put in wine and vinegar while it was being transported from Holland to Hainault, because it would otherwise have spoiled.[35] In the household of the Count of Holland, however, so much wine-vinegar and beer-vinegar was stocked in 1401,[36] that it cannot all have been intended for human consumption. I conjecture that fish was to be preserved in it, but I have not yet found any recipes for this.

A perishable article that demanded preservation, especially in a cattle-raising country such as the Netherlands, was milk. The latter was hardly, if at all, consumed as a drink by adults, but it was eaten by them poured over porridge, after this had been cooked in water. In the household of the Bishop of Utrecht about 22 litres of milk were used per week in 1377/8; at the toll-house of Lobith, in 1426/7, about 6 litres.[37] I do not know how many persons constituted the episcopal household; that of the toll-house, when there were no extra labourers or guests, comprised about sixteen persons. In this period cattle were bred not for milk but primarily for meat, so that their milk production was much less than it is today. Still, there must have been a surplus of milk for conservation after the portion used for porridge had been subtracted. Butter and cheese were made from this because these articles could be kept and stored.

Butter, fresh or salted in barrels, was an important ingredient in the Dutch kitchen[38]—in contrast to that of the countries around the Netherlands, in which butter was of little consequence. There, lard, bacon or oil was used in preparations which, in our country, were made with butter. Cheese could be kept even longer and was also more easily transported. Already in this period, various kinds were known: in the Abbey of Egmond in 1387/8, for instance, cow's cheese, sheep's cheese and *Harmersche* (or *Harnescher*) cheese was eaten. Was the latter perhaps a compressed cheese with a hard rind, in contrast to the usual cream-cheese? Considering the large quantities which were bought and stored of this cheese, I think it must have been not goat's cheese but a cow's cheese with good keeping qualities.[39] Also, the city of Dordrecht, when provisions were bought for two military compaigns in 1286, made various purchases of cheese.[40]

That eggs also were preserved, for instance by being laid in chalk, does not, however, appear from the accounts. In normal households in peace-time they were consumed in very large quantities, probably fresh;[41] and in periods of war, during military campaigns, one had to do without them.

CONCLUSIONS

When we review the evidence, it appears that the following methods of preservation were known in the medieval Netherlands: drying—of subtropical fruits such as dates, figs, raisins and currants, and of peas and fish (dried plaice, basket-herring and stockfish); smoking—of fish, such as herring (which then becomes red herring) and eel, and, probably, also of pork (ham and bacon); butter and cheese-making—from milk; and, last but not least, salting—of butter, meat and fish. As far as salted meat is concerned, this was primarily pork and, for the rest, beef and mutton, although the latter must usually have been eaten fresh. As for fish, salted herring was especially consumed, but also salted cod and eel. Besides this, many fresh-water fish were eaten in the Netherlands, caught by the consumers themselves or bought by them on the market. The preservation of fruit in sugar does not seem to have been practised; and of the pickling of fish in vinegar there is little direct evidence.

The role of preserved food in the medieval Low Countries seems to be as follows: during military campaigns, preserved food, especially salted fish, was consumed in large quantities; in peace-time, preserved food was eaten alongside fresh food in all seasons. There is evidence of more intensive food preservation processing in the autumn, when pigs and oxen were slaughtered. Other cattle could be kept alive, however, until just before their consumption, so that, for these, there was no special slaughtering-month and, consequently, less meat to be salted than one might at first suppose. After salting, drying was probably the most-used method of preservation (for subtropical fruits, pulses and fish), and smoking, too, was not unimportant (especially for herring, eel, ham and bacon). Finally, one should not underestimate the role of cheese-making, a traditional speciality of the Netherlands.

NOTES

General Note on Weights and Measures

Some weights and measures, according to J.M. Verhoeff, *De oude Nederlandse maten en gewichten* (Amsterdam 1983) are as follows:

a *molder* was about 136 litres;

1 *loop* = 4 *schepel*; the Utrecht salt*loop* was the equivalent of 184 litres;

1 *achtel* = 1/8 barrel and was in Egmond the equivalent of about 40 litres;

1 *spint* = 1/4 *schepel* or *achtel* = 1/16 *mud*; the *mud* of peas in Utrecht was the equivalent of about 120 litres.

a *tael* comprised about 200 fish; a *miese* and a *stroe* each about 500 fish;

a barrel or cask of herring contained about 900 fish;

1 *last* herring or red herring = 12 casks, i.e. about 11000 fish.

1 *aem* = 1/6 *voeder*, and a *voeder* or barrel was on the average (depending on local differences) about 900 litres;

1 *aem* = 64 *stoop* = 128 *mengel*;

1 *aem* = 40 *take* = 25 *verdel* = 100 *kwart*;

an *aem* was in Utrecht the equivalent of *ca* 170 litres, a *take* therefore was *ca* 4.25 litres;

an *aem* was at Lobith the equivalent of *ca* 136 litres; a *kwart* therefore was *ca* 1.3 litres.

a barrel or cask butter contained *ca* 160 kilo;

a *cop* was *ca* 600 grams; a *maat* butter was probably 1/4 *cop*, therefore *ca* 150 grams; a *waag* of cheeses had the weight of *ca* 180 lbs.

1. The sources used are especially:

Ch.M. Dozy (ed), *De oudste stadsrekeningen van Dordrecht*, 1284-1424, Werken Historisch Genootschap, 3rd series nr 2, 's Gravenhage 1891 (abbreviated as: Dordrecht).

H.G. Hamaker (ed), *De rekeningen der grafelijkheid van Holland onder het Henegouwsche Huis*, 3 volumes, Werken Historisch Genootschap, 2nd series nrs 21, 24, 26, Utrecht, 1875/6/8 (abbreviated as: Hamaker).

H.J. Smit (ed), *De rekeningen der graven en gravinnen uit het Henegouwsche Huis*, 3 volumes, Werken Historisch Genootschap, 3rd series nrs 46, 54, 69, Amsterdam 1924 and Utrecht 1929-1939 (abbreviated as: Smit).

J.P. Six van Hillegom (ed), *Keuken-rekening van de grafelijkheid van Holland en Zeeland, 1401*, in Kronijk van het Historisch Genootschap, 8 (1852), pp 126-149 (abbreviated as Keuken-rekening 1401).

S. Muller Fz. (ed), *De Registers en rekeningen van het bisdom Utrecht*. 1325-1336, 2 volumes, Werken Historisch Genootschap, 2nd series nrs 53, 54, Utrecht 1889 - 's-Gravenhage 1891; information about food in volume I pp 1-341, Schuldregister van den klerk Hubert van Budel, 1325-1330, and pp 345-403, Huishoudregister van den klerk Hubert van Budel, 1332; and further in volume II, Inleiding, especially on pp xxxiii-lv (abbreviated as: Hubert van Budel).

J.P. Vermeulen (ed), *Bisschoppelijke rekening van 1377-1378*, in Codex Diplomaticus Neerlandicus, Werken Historisch Genootschap, 2nd series nr 2, Utrecht 1853, pp 252-464 (abbreviated as: Bishop 1377/78).

'Accounts of the customs officer at Lobith for the years 1426/1427 and 1427/1428' (unpublished), Rijksarchief in Gelderland at Arnhem, Hertogelijk Archief afdeling A nr 729.

Idem for the year 1428/1429, same archives, afdeling A nr 730.

G.R. Bosscha Erdbrink (ed), *Het 'Keuckenboeck' van de tollenaar van Lobith 1428/1429*, Fontes Minores Medii Aevi 18, Groningen 1979. (abbreviated as: Lobith 1426/27, 1427/28, 1428/29).

J. Hof (ed), Egmondse kloosterrekeningen uit de XIVe eeuw, Fontes Minores Medii Aevi 17, Groningen 1976 (abbreviated as: Egmond).

' "Opkoemen des scaffeners" of the House of the Teutonic Knights in Utrecht for the years 1377/1378 and 1378/1379' (unpublished), Archives of the

'Duitse Huis' in Utrecht, inv nr 646-2 (abbreviated as: Teutonic Knights 1377/78 and 1378/79).

2. Most of the 14th century accounts of the lords of Blois have not yet been published. About the food mentioned in these, see: C.J. de Lange van Wijngaarden, *Geschiedenis der Heeren en beschrijving der stad van Goudq* I, Amsterdam - Den Haag 1813, pp 669-84.

3. Bishop 1377/78 p 461, 'dat vleysch dat jnden kupen was', p 463, 'een draghel tonne runtvleysch'.

4. Dried plaice, basket-herring, red herring, stockfish:
    -   Egmond p 55, account-year 1387/88, purchase of 2 *tael* plaice.   Since a *tael* comprised about 200 pieces, these 400 plaice must have been dried.
        Egmond p 106, in the same account-year, 3 *miese* red herring were bought; p 126 once more 3 *miese* red herring.
    -   Hubert van Budel, I p 353, 5 June 1332, purchase of a basket of plaice; I p 359, 3 July 1332, 3 baskets of plaice and dab ('mandis scullorum et siccorum');
        I p 360, 10 July 1332, dried plaice.
    -   Bishop 1377/78 p 427, 30 and 31 July 1378, dried plaice; p 429, 23 August 1378, dried plaice (and more examples of this).
    -   Keuken-rekening 1401, pp 138-9, May, 10*tael* and 7 *tael* basket-herring; pp 140-1, end of May, 14 *tael* and 7 *tael* basket-herring;
        pp 142-3, June, 10 *tael* and 6 *tael* basket-herring and 1 1/2 *tael* plaice;
        p 145, end of June, 6 *tael* basket-herring;
        pp 146, 148-9, July, 12 *tael* and 8 *tael* and 14 *tael* basket-herring.
    -   In Lobith in 1426/27 fol 46, the customs officer still had 443 stockfish in his stores and he received 960 more of these. Of these 1403 fish, 982 were consumed in this account-year at the expense of the duke (who stayed at the toll-house that year for about 19 weeks) and 172 fish at the expense of the customs officer and his staff. So 249 fish were left to be consumed in the following account-years.
    -   Lobith 1426/27 fol 51, received from the steward; 4*stroe* red herring. In the week of 13-19 April 1427, two more *stroe* of red herring were purchased.
    -   Lobith 1426/27 fol 77, in the week of 6-12 July 1427 basket-herring were purchased:
        Lobith 1427/28 fol 131, in the week of 13-19 July 1427 the same, as also fol 132 verso, in the week of 27 July-3 August 1427:
        fol 133 verso, 10-16 August 1427:
        fol 149, 9-15 May 1428. Lobith 1428/29 p 35 the same, in the week of 3-12 July 1429 basket-herring were purchased.
    -   Lobith 1427/28 fol 145 verso, in the week of 7-13 March 1428 a basket of plaice was purchased.

5. Cowherds:
    -       Hamaker, II pp 163-4 and 420-3, around Palm Sunday 1345, 95 and 20
    live cattle were purchased for the war against the West-Frisians. The animals
    were driven to the northern part of North-Holland and taken care of by servants.
    -       Teutonic Knights 1377/78 pp 27, 29, 41: in October 1377 oxen were
    fetched from Holland and in November and December 1377; and in November
    and December 1377 and in May and September 1378 wages were paid for the
    herding. Item wages for herding 1378/79 pp 59,88: November 1378 and June
    1379.
    -       Lobith 1426/27 fol 48 verso: 8 live oxen were brought to Lobith in the
    middle of October and fed at the expense of the duke until they were, gradually,
    slaughtered. At Easter, 3 were still alive. Besides those, 44 oxen were
    slaughtered and preserved in October and November; see note 8.
6. Skins and fleeces of cattle and sheep sold, and fat for candles:
    -       Hamaker I, p 15-16, the year 1317; II, pp 3-6, 1343/44; pp 110-12,
    1344/45.
    -       Bishop 1377/78 p 265, receipts for cattle-skins sold.
    -       Teutonic Knights 1377/78 p 26, receipts from the sale of cattle-skins and
    sheep's fleeces; 1378/79 p 58, the same of cattle-skins.
    -       Keuken-rekening 1401 p 129-131, receipts from skins, fleeces and *roet*,
    that is to say fat for candles.
    -       In Lobith 1426/27 fol 50 recto and 1427/28 fol 125 recto and verso, the
    tallow of the oxen slaughtered there was made into candles; 1428/29 p 39 wages
    were paid for making these candles. Since the supply of tallow was not
    sufficient, additional candles were purchased.
7. Bishop 1377/78 p 287, 25 October, a sheep and half a cow; 26 October, half a
    cow;
      p 288, 1 November, 3 sheep;
      p 289, 2 November, a wether; 3 November, 3 wethers, half a cow and a hog's
    back; 5 November, a cow;
      p 290, 8 November, half a cow, 2 sheep and a ham;
      pp 290-1, 9 November, half an ox and 3 sheep;
      p 291, 10 November, half an ox, beef and mutton; and so on throughout the
    year, except during fasts and during the periods of the bishop's residence at one
    of his castles outside the city.
8. Lobith 1426/27 fol 32 verso - 33 recto, in October ten barrels of wine were
    purchased to be converted into meat-tubs for the salting of oxen, 44 oxen were
    purchased for slaughter, and 33 *molder* salt. The slaughter took place in the
    weeks of 20-26 October and 3-9 November 1426.
9. Bishop 1377/78 pp 265 and 426, 7 cows slaughtered and preserved, for which 2
    mud salt was necessary. The same, p 463, at the castle Stoutenburg in that
    same year 10 small cows and a bull, as well as 32 sheep and lambs were
    received above and beyond this.

10. Sheep in large numbers:
    - Hamaker, II p 5, in the account-year 1343/44, 456 lambs were consumed, of which 40 at Easter and 181 at Whitsun.
    - For sheep in the bishop's household, 1377/78, see above note 7.
    - Teutonic Knights 1378/79 p 85, May 1379, purchase of 49 sheep.
    - Egmond pp 109-10, in the account-year 1387/88, 6 sheep and 2 rams were bought on All Saints Day, in the week of Easter 4 rams, and on Ascension Day 13 rams.
    - Keuken-rekening 1401 pp 129-30, around Whitsun 16 sheep were consumed in the household of the Count of Holland, in July, 107 sheep, and in the rest of that year another 122 sheep.
    - Lobith 1426/27 fol 40 verso, in the beginning of August 1426, 65 wethers were purchased for storage;
    fol 47, the customs officer received an additional 37 sheep from the steward. In that account-year 129 1/4 sheep were consumed at the expense of the duke; at the customs officer's expense 24 1/2 sheep - together more than he received that year. In that same year, three sheep died through illness, a fact which makes it probable that the others were kept alive in a herd until just before their consumption.
    - Lobith 1427/28 fol 120 verso, purchase of 55 wethers for storage.
    - Lobith 1427/28 p 44, received from the steward 25 sheep.
11. See above, note 6.
12. See the following note.
13. Bacon and other parts of the pig:
    - Hamaker, II p 419, purchased for the war against the Frisians in August-September 1345, 77 joints and 15 hams;
    Hamaker, III p 449, for the siege of Utrecht in June-July 1345 purchase of 7 pigs, 2492 *vlecken*, 14 *baken*. *Vlecken* were halves of slaughtered pigs without heads and ham, *baken* were sides of bacon.
    - Hamaker, III pp 445-47, in May 1345 for the household of the Count of Holland at The Hague purchase of 3 *vlecken*, 1 *bake*; 8 *vlecken*, a boar (male pig), 11 pigs and 5 *baken*; 4 *vlecken*, 2 pork joints, 12 hams, one ton brawn.
    - Egmond p 126 in the account-year 1387/88 one side of bacon was purchased; pp 55 and 110, twice 4 vlecken.
    - Lobith 1426/27 fol 40 verso, 90 sides of bacon purchased; fol 44 verso, 101 pigs received, of which three were ill and one died. Of the healthy pigs, 43 were gradually slaughtered to be roasted for the duke and the remaining 54 pigs were made into 108 sides of bacon. On fol 49 the fodder for the 43 pigs was accounted for.
    - Lobith 1427/28 fol 122 verso, purchase of 66 pigs, of which 6 to be roasted and 2 ill animals. The remaining 58 pigs were converted into 116 sides of bacon.

14. Teutonic Knights 1377/78 p 27, October 1377 purchase of 10*loop* salt.
    November 1377 purchase of a boar for the kitchen, and December 1377, 8 pigs;
    -        Teutonic Knights 1378/79 p 49, October 1378, purchase of 8 pigs and 11
    loop salt, and November 1378 purchase of 2 boars for the kitchen and 4 fatted
    cows.
15. Sausages do appear in Bishop 1377/78 p 461: 'ende die worsten van V verken',
    but this is an exceptional case.
16. C.A. S[errure] (ed), *Keukenboek uitgegeven naar een handschrift der vijftiende
    eeuw*, Maatschappij der Vlaamsche Bibliophilen (Gent 1872);*Een Notabel
    Boecxken van Cokeryen*, facsimile-edition with the title 'Het eerste
    Nederlandsche gedrukte kookboek (Brussel, Thomas van der Noot,*ca* 1510)', 's-
    Gravenhage, 1925.
17. Keukenboek Vlaamsche Bibliofilen, p 1, nr III (stuffing), p 2 nrs IV, V, VI
    (used for pigs feet, lobster and apples). Other pork recipes: p 3 nr IX (pigs feet
    in jelly), pp 5-6, nr XX (swan's neck filled with pork) and p 12 nr XII (suckling
    pig in peppersauce).
18. Countess Johanna of Holland-Hainault consumed in 1326/27 at her court in
    Hainault 19 1/2 deer, of which 17 were fresh and 2 1/2 salted, Smit, I p 290.
    In June 1344 a huntsman provided the kitchen of the Count of Holland-
    Hainault with two sides of a deer and a leg, Hamaker, III p 313; and in April
    1345 with a deer and a half a wild boar, Hamaker, III p 443.
    In September 1344 he received from the huntsmen at his court at The Hague
    260 rabbits, 5 deer and 188 partridges, Hamaker, III p 318.
19. Chickens, capons, ducks and geese:
    -        Hubert van Budel, I pp 351-52, 31 May 1332, 18 geese, 7 capons, 8
    cocks and 14 chickens;
    I p 373, 9 September 1332, 11 geese and 43 chickens; in this account fowl is
    mentioned almost every week, but only very seldom the quantity is specified.
    -        Bishop 1377/78 p 283, 11 October 1377, 6 geese and 14 ducks and 8
    chickens; 12 October 1377, 10 ducks and 6 chickens;
    13 October, 5 geese and 6 ducks;
    p 284, 15 October, 12 ducks and 17 chickens; p 285, 18 October, 6 geese and
    12 ducks; 19 October, 10 ducks, 20 October 7 geese and 10 ducks;
    p 286, 22 October, 12 ducks and 2 chickens; p 287, 25 October, 12 ducks; 26
    October, 18 ducks and 6 chickens; 28 October, 14 ducks and 2 chickens;
    pp 287-8, 29 October, 16 ducks and 8 chickens;
    p 288, 1 November, 6 geese and 12 ducks and 4 capons and 4 chickens.
    Etcetera, except during the fasting periods. These were not the only fowl that
    were eaten at this court: plovers and wood-cocks and small birds were also
    consumed regularly.
    -        Lobith 1426/27 fol 58, 22-28 September 1426, 16 persons, 2 geese, 6
    ducks and 11 chickens;

fol 58 verso - 59 recto, 29 September-5 October, 16 persons, 3 geese, 15 ducks, 14 chickens;

fol 59 verso, 6-12 October, 21 persons, 2 geese, 21 ducks, 3 chickens; fol 60, 13-19 October, 17 persons, 4 geese, 6 chickens, etcetera.

fol 66, 2-8 February 1427, 18 persons, 22 chickens and a capon;

fol 66 verso, 9-15 February, 17 persons, 36 chickens and a capon. Etcetera, except during fasting-periods.

20. Herons and bitterns:
- Hubert van Budel, I p 368, 20 August 1332, a heron and a bittern purchased;

p 373, 9 September 1332, 2 bitterns, and a heron and a bittern.

- Egmond p 110, in the account-year 1387/88 12 herons received as gifts, for which the messenger who brought them was given a reward.

- Keuken-rekening 1401 p 139, 19 May, 19 herons;

p 141, 28 May, 5 herons and 3 bitterns;

p 142, beginning of June, 7 herons and 29 bitterns;

p 146, beginning of July, 22 bitterns;

p 147, 16 July, 4 bitterns;

p 148, second half of July, 14 bitterns;

p 149, July, 44 herons.

21. On the technique of gutting and curing herring and the various sorts of herring sold on Dutch markets, see: Richard W. Unger, 'The Netherlands herring fishery in the late Middle Ages', in *Viator, Medieval and Renaissance Studies,* 9 (1978) pp 335-56.

22. Herring:
- Hubert van Budel, I p 359, 3 July 1332, small salted herrings ('de parvis allecibus'; 'allec' was the moist salted herring);

p 364, 31 July, 200 salted herrings;

p 367, 12-16 August, 300 salted herrings, eel and other fish;

p 368, 21 August, 'pro harenc' (obviously not the usual sort; perhaps fresh?);

p 375, 14 September, salted herring. Etcetera.

- Bishop 1377/78 p 359, 16 January 1378, 100 herrings and a tael green herring (*ca* 200 fish);

pp 436-7, 19 and 24 November 1377, green herring. Etcetera.

- Egmond pp 105 and 107: a cask herring and two casks herring purchased, in the account-year 1387/88.

- Keuken-rekening 1401, basket-herring: see above, note 4.

- Lobith 1426/27 fol 40 verso, six casks herring purchased.

- Lobith 1427/28 fol. 124, 26 casks herring received and bought, of which almost 15 casks were consumed at the expense of the duke and almost 10 casks at the expense of the customs officer; a small supply remained. For basket-herring at Lobith, see note 4.

23. Lobith 1426/27 fol 55 verso, in the week of 1-7 September 1426, 4 stockfish were consumed;

    fol 57, in the week of 15-21 September 10 stockfish;

    fol 58, in the week of 22-28 September 12 stockfish. Etcetera. Only in the weeks after Easter and Whitsun was stockfish omitted from the menu.

24. Hamaker, II pp 168-172, preservation of cod for the campaign of 1345.

25. Hamaker, II p 172, 10 *last* red herring for the campaign against the Frisians;

    Hamaker, II pp 415-17, 45 casks eel and more than 45 1/2*last* red herring for the same campaign .

26. Hamaker, III pp 447-50, purchase of provisions for the siege of Utrecht;

    p 449, total of the meat was 1763 cattle, 267 live animals + 990 slaughtered ones = 1257 sheep, 7 live pigs, 2492 *vlecken*, 14 *baken*;

    p 450, total of the fish was 29 1/2 *last* + 8 *miese* red herring, 12 casks herring, 9 casks eel, 3183 salmon, 3500 + 7325 salted cod, 9 casks haddock which were thrown away (because of spoilage), 2 porpoises.

    Hamaker, III pp 431-2, the siege was lifted on 22 July. See on this siege: J.E.A.L. Struick, 'Om ijdel woorden wille', in *Jaarboek Oud-Utrecht 1981*, p 35-60.

27. Egmond p 40, account-year 1344/45, 15 *tael* eel;

    Lobith 1427/28 fol 149, half a cask of eel.

28. Pulses:

    -    Egmond p 40 account-year 1344/45, half a barrel and a whole barrel peas purchased;

    pp 107-8, 1387/88, 24 1/2 *achtel* peas purchased, i.e. 3 barrels and half an *achtel*;

    p 126, 3 *achtel* peas purchased, and 2 *achtel* beans for the poor.

    -    Bishop 1377/78 p 284, 16 October 1377, a *spint* peas purchased; p 416, 19 March 1378, 2 *mud* chick peas and a *mud* green peas purchased.

    -    Teutonic Knights 1377/78 p 28 , in January 1378 [1.5 *mud*] beans purchased, and in April 1378, 3 *mud* beans;

    -    Teutonic Knights 1378/79 p 60, in January and April 1379 peas and mustard seed purchased.

    -    Lobith 1426/27 fol 58, 22-28 September 1426, 16 persons, purchase of 1/2 schepel peas;

    fol 58 verso - 59 recto, 29 September-5 October, 16 persons, 1/2 *schepel* peas;

    fol 59 verso 6-12 October, 21 persons, 1/2 *schepel* peas;

    fol 60, 13-19 October, 17 persons, 3/8 *schepel* peas. Etcetera.

    fol 66, 2-8 February 1427, 18 persons, 3/8 *schepel* peas;

    fol 66 verso, 9-15 February, 17 persons, 3/8 *schepel* peas. Etcetera.

29. Subtropical fruits:

    -    Egmond p 42, account-year 1344/45, a 'coppel froyts', i.e. a basket of figs and a basket of raisins purchased, and 20 lbs almonds;

p 106, 1387/88, a 'coppel' figs and raisins, and 16 + 14 = 30 lbs almonds purchased.

- Bishop 1377/78 p 308, 18 March 1378, for the kitchen 2 lbs figs and 2 lbs raisins purchased; 19 March, 4 lbs almonds;

pp 308-9, 21 March, 1lb figs and 1lb raisins and 1.5lb almonds;

p 336, 17 March 1378, for the chamber of the bishop 2lbs figs and 2lbs raisins purchased;

p 337, 19 March, for the chamber 2lbs figs and 2lbs raisins; 20 March 1lb figs; 21 March, 2lbs figs and 1lb raisins.

- Lobith 1426/27 fol 51, received 3 baskets of figs.

- Lobith 1427/28 fol 126, received 4 baskets of figs and 3 baskets of raisins, which were consumed at the expense of the duke.

- Lobith 1428/29 p 47, received 4 baskets of raisins, of which 2 were spoiled.

30. Karl-Heinz Knorzer and Gustav Muller, 'Mittelalterliche Fäkalien-Fassgrube mit Pflanzenreste aus Neuss', in *Rheinische Ausgrabungen 1, Beiträge zur Archäologie des Mittelalters*, Köln-Graz 1968 (Beihefte der Bonner Jahrbücher, Bd. 28) pp 131-169.

- L.M. van den Brink in H.L. de Groot, Dorstige Harthof en omgeving, in 'Archeologische en bouwhistorische kroniek van de gemeente Utrecht over 1983', *Maandblad Oud-Utrecht* 57 (1984) pp 108-13.

- M.A.G. Sengers, 'Keukenstraat 10 - 42 /Schalkwijkstraat 1a-1b, Botanisch onderzoek', in 'Archeologische en bouwhistorische kroniek van de gemeente Utrecht over 1985', *Maandblad Oud-Utrecht* 59 (1986) pp 164-5.

- *idem*, 'Lange Lauwerstraat 3-13, Botanisch onderzoek', in *ibid* pp 180-1.

31. *Keukenboek Vlaamsche Bibliophilen* p 2 nr VI, stuffed apples (filled with a mixture of pork, hard boiled eggs and spices); and p 5 nr XVI, apple sauce for the fasting periods with fish liver and almond milk. *Notabel Boecxken*, fol E II verso - E III recto, apple pie;

fol E III recto, apple sauce;

fol E III verso - E IV recto, apple turnovers. Universiteitsbibliotheek Gent, 'ms nr 476, cookery-book from the 16th century' (unpublished),

fol 43 verso, 'Om peren inden ijpocras'.

32. Bishop 1377/78 pp 330-5, pears for the bishop's chamber in the autumn of 1377, mentioned between the wax candles and the tallow candles;

pp 333-6, April-May 1378, apples for the bishop's chamber;

pp 339-40, August 1378, pears and nuts.

33. Teutonic Knights 1377/78 pp 32, 33, 34, apples for the cellar purchased in November 1377, February, April and May 1378;

ibid 1378/79 pp 62, 63, 64, November and December 1378, and April 1379.

34. Wine and beer:

- Hubert van Budel, I p 346, 22 May 1332, purchase of a barrel of beer and 50 *take* wine and 10 *take* red wine;

p 357, 27 May-21 June 1332, 'Dominus bibit 1 vas rubei vini' (the lord drank one barrel of red wine);

p 361, 2-13 July 1332, purchase of 2 barrels of red wine, a *fust* white wine of 5 *aem* - 10 *take*, and 2 *aem* + 10 *take* white wine, and 10 barrels of beer;

p 367, 12-16 August 1332, 30 *take* wine purchased and brought to the episcopal castle at Ter Horst;

p 387, 26 September 1332, 44 *take* new wine;

p 398-400, 15-29 November 1332, purchase of 80.5 *take* and 1 *mengel* wine.

- Egmond p 104, account-year 1387/88, purchase of 91 barrels of Hamburger beer, a barrel of Delft beer, 42 1/2 barrels of hop beer and 12 barrels of barley-beer;

p 105, purchase of 6 *aem* and 13 *stoop* wine.

- Lobith 1426/27 fol 39 recto - 40 recto, purchase of 51 *voeder* and 1/2 *aem* wine, of all kinds.

- Lobith 1427/28 fol 122 verso - 123 recto, left over from last year 4 *voeder*, 3 *aem* and 22.5 *verdel* wine, and also received (including the old supply) 17 *veoder*, 3 *aem* and 11 *verdel*.

- Lobith 1428/29 pp 41-2, left over 3 *voeder*, 3 *aem* and 5 *verdel* wine; and also received up to a total (inclusive) of 9 *voeder*, 1 *aem* and 17 *verdel* wine, and a small barrel of wine of 15 *verdel*.

35. De Lange van Wijngaarden, pp 679-80.

36. Egmond p 39, account-year 1344/45, purchase of 2 barrels of beer-vinegar;

p 41, a barrel of wine-vinegar;

p 105, 1387/88, half a barrel of wine-vinegar.

- Hamaker, III p 321, a cask of wine-vinegar purchased in December 1344.

- Keuken-rekening 1401 pp 136-7, purchase of 6 barrels of beer-vinegar and 6 *karlagelen* wine-vinegar;

p 138, May, 5 barrels of vinegar;

pp 142-3, first half of June, a small cask of vinegar and a cask of beer-vinegar and half a barrel of wine-vinegar;

p 144, second half of June, 2 *karlagelen* of wine-vinegar and a cask of beer vinegar;

p 146, beginning of July, a cask of wine-vinegar;

p 148, end of July, a small barrel of wine-vinegar.

37. There was no fixed rate of milk consumption per week, but the bishop's household, for instance, consumed an average of 5 *take* a week, and the customs officer's household at Lobith, 4 *kwart* a week. During fasting periods, especially the forty days before Easter, milk was omitted entirely.

- Bishop 1377/78 p 283, Monday 12 October 1377, 2 *take* milk;

p 285, Saturday 17 October, 2 *take* milk, Monday 19 October, 1 *take* milk;

p 286, Wednesday 21 October, 2 *take* milk, Friday 23 October, 2 *take* milk;

p 287, Wednesday 28 October, 1 1/2 *take* milk;

p 288, Friday 30 October, 1 1/2 *take* milk;

p 289, Monday 2 November, 4 *take* milk, Wednesday 4 November, 4 *take* milk;

p 290, Friday 6 November, 2 *take* milk;

p 291, Monday 9 November, 1 *take* milk;

p 292, Monday 16 November, 4 *take* milk, Wednesday 18 November, 3 *take* milk. Etcetera.

-      Lobith 1426/27 fol 58, 22-28 September 1426, 16 persons, 8 *kwart* milk;

  fol 58 verso-59, 29 September-5 October, 16 persons, 4 *kwart* milk;

  fol 59 verso, 6-12 October, 21 persons, 4 *kwart* milk;

  fol 60, 13-19 October, 17 persons, 4 *kwart* milk. Etcetera.

  fol 66, 2-8 February 1427, 18 persons, no milk;

  fol 66 verso, 9-15 February, 17 persons, 2 *kwart* milk;

  fol 67 verso, 16-22 February 23 persons, 4 *kwart* milk;

  fol 68, 23 February-1 March, 23 persons, 8 *kwart* milk;

  fol 68 verso, 2-4 March, 28 persons, 4 *kwart* milk;

  fol 69-71, 5 March-19 April, Lent, no milk;

  fol 71, 20-26 April, 20 persons, 2 *kwart* milk;

  fol 71 verso, 27 April-3 May, 23 persons, no milk;

  fol 72, 4-10 May, 23 persons, 5 kwart milk. Etcetera.

38. Butter:
  -      Hubert van Budel, I p 364, 1 August 1332, 5 *cop* butter; p 373, 7 September 1332, a *cop* butter.
  -      Bishop 1377/78 p 283, 9 October 1377, 12 *maat* butter; 10 October 16 *maat* butter;

    p 284, 14 October, 8 *maat* butter; 16 October, 15 *maat* butter and 6 lbs of old butter;

    p 285, 17 October, 12 *maat* butter and 10 lbs of old butter; 19 October, 4 *maat* butter; 20 October, 4 *maat* butter;

    p 286, 21 October, 9 *maat* butter and 10 lbs of old butter; 22 October, 4 *maat* butter; 23 October, 12 *maat* butter and 10 lbs of old butter; 24 October, 12 *maat* butter and 10 lbs of old butter;

    p 287, 27 October, 4 *maat* butter and 10 lbs of old butter; 28 October, 8 *maat* butter;

    p 288, 30 October, 12 *maat* butter and 6 lbs of old butter;

    p 289, 2 November, 9 *maat* butter. Etcetera.
  -      Egmond p 108, account-year 1387/88, purchase of 4.5 barrels of butter.
  -      Lobith 1426/27 fol 45 verso, received from the steward 14 barrels of butter and 1.5 *kwart* barrel of butter.
  -      Lobith 1427/28 fol 123 verso, 5 barrels of butter purchased.
  -      Lobith 1428/29 p 43, received half a barrel of butter; alongside this salted butter, fresh butter continued to be bought.

39. Cheese:
  -      Egmond p 55, account-year 1387/88, purchase of 10 cow's cheeses;

p 97, 500 sheep's cheeses;

pp 108-9, a *waag* 'Harmersche' (or 'Harnescher') cheeses, 475 sheep's cheeses and 51 cow's cheeses.

- Also at Lobith a considerable amount of cheese was eaten: Lobith 1426/27 fol 40 verso, 5 waag cheeses purchased, comprising altogether 90 'kanter'-cheeses;

fol 46, in addition 125 cheeses received.

- Lobith 1427/28 fol 120 verso, 6 waag cheeses purchased, comprising 90 cheeses.

- Lobith 1428/29 pp 43-4, left over from previous years 78 cheeses and received in addition 66 cheeses, which add up to 144 cheeses; of these, 28 were consumed at the duke's expense and 70 at the expense of the customs officer; 15 were spoiled, so that 31 cheeses remained.

40. Dordrecht p 66, 69, 73, cheese for the military campaigns of 15 September and 31 October 1286.

41. Eggs in the accounts, only those cases in which the quantity is specified:
    - Hubert van Budel, I p 350, 29 May 1332, 200 eggs;

    p 364, 1 August 1332, 400 eggs;

    p 394, 10 November 1332, 100 eggs;

    p 399, 20 November 1332, 100 eggs.

    - Bishop 1377/78 p 283, 10 October 1377, 200 eggs;

    p 284, 14 October 1377, 200 eggs; 16 October, 200 eggs;

    p 285, 17 October, 300 eggs;

    p 286, 23 October, 200 eggs; 24 October, 200 eggs;

    p 288, 30 October, 300 eggs;

    p 289, 2 November, 100 eggs. Etcetera.

    - Lobith 1426/27 fol 58, 22-28 September 1426, 16 persons, about 60 eggs;

    fol 59 verso - 59 recto, 29 September-5 October, 16 persons, 175 eggs;

    fol 59 verso, 6-12 October, 21 persons, 100 eggs;

    fol 60, 13-19 October, 17 persons, 125 eggs. Etcetera.

    fol 66, 2-8 February 1427, 18 persons, 250 eggs;

    fol 66 verso, 9-15 February, 17 persons, 300 eggs;

    fol 67 verso, 16-22 February, 23 persons, 200 eggs;

    fol 68, 23 February-1 March, 23 persons, 250 eggs;

    fol 68 verso, 2-4 March (Carnival), 28 persons, *ca* 200 eggs;

    fol 69-71, 5 March-19 April, Lent, no eggs;

    fol 71, 20-26 April, 20 persons, 400 eggs (Easter on 20 April);

    fol 71 verso, 27 April-3 May, 23 persons, 100 eggs;

    fol 72, 4-10 May, 23 persons, 150 eggs. Etcetera.

BAKING, BREWING, AND FERMENTING OF THE GRAPE-MUST: Historical Proofs of their Connections

Rudolf Weinhold

In this paper I want to report some observations and reflections derived from matters pertinent to this conference. I shall deal with one aspect of the relations between baking, brewing, and fermenting of the grape-must; or, more precisely, with an apparatus used in these processes.

The apparatus has preserved its functions, and probably its elementary form too, for a long time; so it could provide circumstantial evidence for the hypothesis that the preparation of dough as well as brewing and fermenting were based on the empirical utilization of a method closely connected with it.

What I am talking about is a long wooden trough, which was either carved out of a tree-trunk or made by joining together some big planks.

Early evidence for such a big vessel has been found in Old Egyptian paintings and reliefs. The earliest come from a tomb of the Fifth Dynasty (i.e. the second half of the third millennium BC). Further pieces of evidence have come to light in necropolises of the Middle and New Kingdom. All occur in texts reporting on the brewing of beer by the ancient inhabitants of the Nile Delta.

Such descriptions usually begin with the kneading, moulding and semi-baking of the so-called beer-breads. These, both in Egypt and in Mesopotamia, represented the initial product of the brewing process. For this purpose raw or pre-germinated barley or spelt-corn was used, after being crushed in a mortar. The loaves made from this material were allowed to cool, then broken into pieces and soaked with water in earthenware or wooden vessels, until they were interfused with moisture like a sponge. This moist substance was then mashed and stirred to produce a thick pulp. The mashing was done by means of pestles or even, quite often, by treading. In the latter case the vessel was not of earthenware; a wooden 'canoe-trough' (*Einbaum*-trough), strong enough to take the weight of one or two people, was used (figure 1). The Old Egyptian paintings copied the natural veining of the material, showing conclusively that it really was a wooden vessel.

When the mass was sufficiently trodden and had assumed the right consistency, it was packed into baskets or wooden troughs and taken to the cellar. There it was turned out into sieves, from which clarified liquid steadily descended into the fermenting tubs below.

It is remarkable that the Old Egyptian illustrations nearly always show brewing and baking in close proximity. The pictures of the two operations are usually separated from each other by no more than a simple line. Sometimes even this line is missing, so that 'the several pictures appear to be scenes of one continuous process taking place in a single workroom. Some reliefs even show the bakery in the middle of the brewery operations'.[1]

Figure 1. Painting from the Tomb of Nakht, 15th century BC, showing a walled trough and the arbour with ropes for the treading men to hold.

Quite evidently, in Old Egypt brewing and baking belonged closely together. On large landed properties the two trades were 'certainly assigned one and the same room as workshop, the employees were under the control of the same overseers, and they could without much ado have been assigned to either activity. The baker could brew beer, and the brewer could bake bread. Part of the brewing process—nearly half of the technical operations involved—anyway fell into the work-province of the baker: from preparing the raw corn up to production of the beer-breads, semi-baked in the oven or in heated baking-forms.'[2]

The close connection of the two occupations is still visible millennia later, from the famous ground plan of the Monastery of St Gall (dating from AD 820). Here also, brewing and baking always took place side by side 'in a kind of double-house'.[3] This being so, it is not surprising that some correspondences of implements are to be observed, especially with regard to the type of trough being used. It is likely that these troughs had, to some extent, a dual function; they could be used for brewing as well as for preparing bread dough. Their sides and bottoms—and this point is decisive—were always colonised by both baking- and brewing-yeasts. Since both were present, they could start and effect the fermentation of the bread dough as well as that of the beer. Advantage had been taken of this characteristic for a long time for the purpose of preparing simple foods so that they would keep for a while.

In the light of this knowledge, we must ask whether the Finnish *kuurna* (analysed in detail by Matti Räsänen[4]) and the North and East

European relations thereof may also be included in this group of vessels. These Finnish *kuurna* are tubs carved out of a tree-trunk, with a semi-circular cross-section and a length of one to three metres (figure 2). Their width and depth vary from 20 to 80 cm. They were usually made from aspen-wood, which could easily be hollowed out, or of sound pine-wood. On both of its narrow ends, the trough was closed by boards fitted in exactly. One of its ends jutted out slightly and was equipped with a hollow groove, which could be closed with a plug. The *kuurna* was placed in a somewhat slanting position—the end with the hollowed groove being the lower—on a special support (a bench or a couple of chairs).

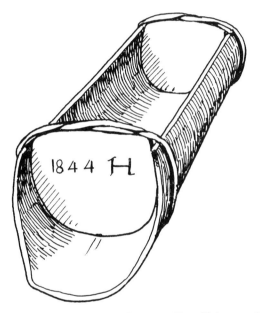

Figure 2. The Finnish kuurna *from M. Räsänen:* Vom Halm zum Fass *(Helsinki, 1975). Drawn from a photograph by Ingrid Lowzow.*

However, the primary function of the *kuurna* differs from that of the brewing-trough previously described. The *kuurna* was used above all for clarifying the wort out of the mash. For this purpose a lattice-like layer of wood shavings was put in the trough, then covered with long blades of rye-straw and oat-straw.

Upon this 'bed', which acted like a sieve, was poured the mash, i.e. the kiln-dried malt of the pre-germinated corn mixed with water. Progressively hotter water was then poured over the mash.

Sometimes the malt was also heated directly in the trough. In that case, after the pre-germinated corn had been poured in, the *kuurna* was filled with water, which was then heated to a high temperature by carefully placing heated stones in it. (This method, by the way, is also known from

Carinthia[5], Styria[6] and parts of Poland[7]; but even in those places it has not survived into the 20th century.)

During this process the enzymes, maltase and diastase, present in the malt convert the substances in the malt, mainly the starch, into the water-soluble sugar maltose and dextrin. Together with other ferments, maltose and dextrin account for the flavour and salubrity of the beer.

The liquid thus produced in the *kuurna* is left to clarify. When it is clear, it is let out through the hollow groove in the lower end of the *kuurna* into another vessel for further fermentation. This liquid is called wort. The layer of wood shavings and straw in the bottom of the *kuurna* serve as a sieve for it. For our purposes it is important to note that these two stages of the brewing both take place in the *kuurna*.

The mash left in the *kuurna* could be soaked with hot water for a second time, the so-called 'second running'. The beer so produced, of secondary quality, was regarded as an everyday drink. This procedure has a parallel in wine-making, where the 'press-wine', wine of the second running, is called 'low wine' in vine-growing households.

The question remains whether it is possible to establish a connection between the processes taking place in the *kuurna* with those involved in the archaic baking- and beer-trough.

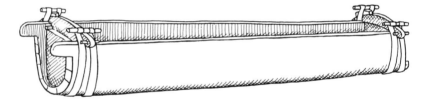

Figure 3. One-piece treading trough, the koruba, *from Podvis, district of Burgas. Bulgarian Academy of Sciences, Ethnographic Institute and Museum, Sofia.*

Our first response must be negative. The function of the *kuurna* is different; it is to clarify and filter off the wort obtained from the mash.

However, there remains the possibility that the warming which the *kuurna* undergoes activates micro-organisms which had already (assuming that the *kuurna* had been in use for some time) colonised its inner sides. These micro-organisms might have included yeast spores, which could have started, or at least helped to start, the first fermentation. In this connection it would be helpful if investigations could be made on the basis of careful biological analysis; these could throw further light on ancient fermentation techniques and the relationships between them.

We stand on firmer ground when we consider the production of must and wine, in respect both of the implements used and the appearance of the yeast spores in the wine lees.

The implements, treading-trough and wine press, are mentioned in a document from the Prussian Government in 1800. In this document the authorities complain that the grapes from the county of Grunberg 'come whole ... without being separated from their stalks ... into the treading-trough or into the wine-press'.[8]

It has been known for a long time that when bits of vine stalk go into the press with the grapes, the colour of the wine is spoiled. It has also been known that spores from the wine-yeast settle down on grapeskins, and we may infer from this that they would also be present on the walls of the treading-trough.

Now this treading-trough was, in shape, closely related to the Old Egyptian and medieval dough-preparing and beer-brewing troughs, and to the Finnish *kuurna*. According to the descriptions handed down,[9] it was a long, not very wide, but deep, tub carved out of a tree-trunk with hatchet and adze in the same way that a canoe (*Einbaum*) was made. Winegrowers in the lower reaches of the Oder and at Guben called it *Keltertrog*, which means 'wine-pressing-trough'.[10]

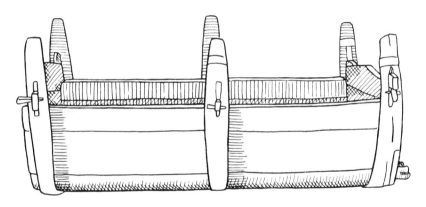

Figure 4. Treading trough made of planks, the korab, *from Ljaskovec, district of Tarnovo. Bulgarian Academy of Sciences, Ethnographic Institute and Museum, Sofia.*

The two names, treading-trough and *Keltertrog*, refer to different functions of the trough. First, the grapes were trodden in it. Secondly, it acted as a basin to collect the must running off from the wine-press. Here we are mainly interested in the first function.

The evidence already cited from 1800 refers unambiguously to this first function. In the troughs which were used in the Silesian wine-growing area the grapes were trodden out and mashed before being pressed out in the wine-press. It is probable that in earlier times the treading process alone was used.

Evidence that this or a similar trough was used in both ways may be found among the older implements of viniculture in south-east Europe. Bulgarian and Romanian wine-growers used similar troughs (whether carved out of tree-trunks or made from big planks) for treading the grapes as well as for collecting the grape-juice from the wine-press.[11] (See figures 3 and 4.) The names they used for the troughs are *korab*, *koruba*, and *lin*. However, it is of great interest that in one Bulgarian dialect, a particular part of the *korab*, namely the flow-pipe, is called *kurna*, almost the same as the Finnish *kuurna*. It would be worth investigating the possibility of a link, perhaps effected through the Volga-Bulgarian of the early Middle Ages, with Finn-Ugrian.

Figure 5. Silenus treading grapes in a trough of joined planks (in the style of the Bulgarian korab). *Painting on an Attic amphora from c. 510 BC. Martin-von-Wagner, University of Würzburg.*

Apparatus of the same type were still being used in the first half of this century in central Greece and in the Racinskij Mountains of Georgia[12].

In Greece, at least, the treading-trough has been in use since antiquity, as demonstrated by paintings on Attic black-figure vases of the 6th century BC (figure 5). These paintings reveal some characteristic details of the construction. The wooden troughs were placed on a stand supported by vertical piles, and had a projecting funnel from which the trodden juice flowed into a collecting vessel. The mashing of the grapes was not done directly onto the bottom of the trough, but in a basket placed in the trough. The liquid flowed off through the wickerwork of this basket, which thus acted as a sieve, holding back the stalks, and probably some of the pips and skins as well.

Figure 6. Silenus treading grapes. The size of the basket in the raised trough is probably exaggerated. Amasis amphora, second quarter of the 6th century BC Martin-von-Wagner, University of Würzburg.

Here too, there is an interesting terminological point. The ancient Greek name for a wine-press, *lenos*, has the primary meaning of 'trough' or 'tub'. In this meaning it precisely describes the apparatus just described. Besides the *calcatorium* (a treading-trough made of stone, in use in the eastern Mediterranean regions and parts of Asia), and the basket of wickerwork, the wooden treading-troughs probably represent the implements originally used for the production of grape juice. One might add that the inevitable incrustation of wine-lees spores on the interior of the sides of the apparatus would have helped to start the alcoholic fermentation through the stage of must towards the result of wine.

Thus this apparatus can be subsumed in a group of wooden vessels which, by their form and function, constitute an indirect proof of the essential technological unity of certain processes: dough-preparing, brewing, and wine production.

REFERENCES

1. E. Huber and M. Philippe, *Bier und Bierbereitung bei den Völkern der Urzeit: 1, Babylonien und Ägypten*, Gesellschaft fur die Geschichte und Bibliographie des Brauwesens, Berlin, 1926, p 50 ff.

2. *Ibid*, p 48.

3. M. Heyne, *Das deutsche Nahrungswesen von den ältesten geschichtlichen Zeiten bis zum 16 Jahrhundert*, Leipzig 1901, p 344.

4. M. Räsänen, *Vom Halm zum Fass*, Helsinki 1975, p 168 ff.

5. R. Dünnwirth, 'Vom Steinbier', in *Carinthia*, vol 95, Klagenfurt 1905, pp 10-19. M. Wutte, 'Eine alte Beschreibung der Steinbiererzeugung', in the same, vol 123, Klagenfurt 1923, p 212.

6. F. Pichler, 'Das Steinbierbrauen in der Steiermark', in *Zeitschrift des Historischen, Vereins für Steiermark*, III, Graz 1962, pp 155-73.

7. A. Maurizio, *Geschichte der gegorenen Getränke*, Wiesbaden 1933 (reprinted 1970), p 111.

8. H. Schmidt, *Geschichte der Stadt Grünberg/Schlesien*, Grünberg 1922, p. 1010.

9. *Ibid*, p 399 ff and 1009 ff. See also W.H. Veith, *Die schlesische Weinbauterminologie in ihren ostmitteldeutschen und gesamtdeutschen Bezügen*, Marburg/Lahn 1966, p 109/112.

10. K. Gauder, *Die Geschichte der Stadt Guben*, Guben 1925, p 493 ff.

11. R. Weinhold, *Vivat Bacchus: Eine Kulturgeschichte des Weinbaus und des Weines*, Leipzig 1975, p 242 ff.

12. *Ibid*, p 130.

Cultural Papers

THE CHANGE IN RURAL FOOD STORING IN A VILLAGE IN NORTH-WEST GERMANY

Christine Aka

In this paper I shall present some of the results of an investigation into the problem of storing food on farms in north-west Germany, undertaken in 1984 as part of my intermediate work at the Volkskundliches Seminar of the WWU Münster. I conducted a number of interviews with women aged between 25 and 90 in the agricultural village of Hegstedt, about 50 km south of Bremen. All the women interviewed ran households on farms. They were questioned about various preserving methods and the storing of provisions generally in the period from 1910 to the present, so as to build up a picture of how food storage has changed.

I do not intend to discuss in detail here the the development of the particular methods of preservation for individual foodstuffs. As Edith Hörandner has said, 'Long-established methods of preservation, ... changed little till the beginning of the 19th century, and were identical or similar throughout the world.'[1] Here we are concerned with looking at the changes in eating habits and in the daily life of the housewife that have been brought about by the transformation of preserving techniques. In a brief survey I shall describe the eating habits and preserving methods of the farming population of Hagstedt, first in the time around 1920, and then in the present day. In conclusion I shall discuss the effects on eating habits and the work of the farmers' wives brought about by the changes in the preservation methods available.

EATING HABITS AND PRESERVING METHODS AROUND 1920

Around 1920 the most important principle in household economy was self-sufficiency. The housewife's aim - and the proof of her abilities - was not to have to buy anything and not to rely on the help of neighbours.

Although, in the village investigated, preserving in glass jars was known from 1910, the daily stew (on the weekday menu all year round in the family of the oldest woman interviewed) did not require foods preserved in this way. It was based on beans, peas, cabbage, carrots and potatoes because these vegetables were easy to preserve otherwise: either by being dried (as in the case of beans and peas); laid in salt (beans and cabbage); or clamped (cabbage, potatoes and carrots). Salted or smoked meat, mostly sausages, was added. Until about 1926 the farmers took the meat to the farm hands who smoked it over open fires in their houses. However, once they too had ceased to use open fires, the farmers had a smoking chamber of their own built in their houses. Apart from very occasional game, and a few chickens and sheep, fresh meat was eaten only on the four or so killing days in the winter, which therefore provided a welcome change. In the interviewee's words: 'How we liked it! We didn't have that again for a whole year.'

Fruit was mostly dried and turned into jam and jelly. But it was only during the harvest that it brought a certain variety to the monotonous diet of the main meals.

Breakfast foods in summer were egg pancakes, and in the winter when the hens laid no eggs because of the cold, groats were eaten. These were cooked from oats and the innards of slaughtered animals. They had to last from one killing to the next, which meant that they were eaten until they had almost gone off.

For the evening meal there was milk soup and black bread all year round. People had this coarse black rye bread, baked at the black bread baker's, as a loaf weighing approximately 15 kg about every three weeks.

The only variation from this fixed menu was provided on Sundays and festivals. For these days, the housewife preserved beef in glass jars which was prepared as a ragout and was also used for making soups. Eating habits in 1920 were still largely based on the annual rhythm of harvests and slaughtering times. Since there was no possibility of preserving food over a very long time or of having a constant fresh supply, the storing of food for the whole year had to be planned in detail far in advance.

EATING HABITS AND PRESERVING METHODS IN 1984

The housekeeping of the younger women interviewed is characterized by the multiplicity of new developments in preservation methods, both in the domestic and industrial spheres, and more general changes, as well as by transformations in agriculture which have occurred mainly since the Second World War. The menu of these families is balanced and full of variety. It includes many different dishes: on the one hand the traditional meals like stew, on the other modern 'trendy' foreign dishes like pizza and spaghetti. The demand for meat is still mostly met by the farm's own livestock, as in the past, but the animals are no longer killed by the farmers themselves, but prepared by the butcher so that the housewife has only to pack the meat in bags and put it in the freezer. Except for beans and peas which are eaten fresh, hardly any vegetables are grown. They are either bought fresh at farms which have specialized in growing vegetables, or else bought tinned, bottled or deep-frozen at the supermarket. Fruit, e.g. strawberries, cherries or apples, is still much grown. They are mostly frozen, or bottled if the freezing capacity is not sufficient. They are also needed for the year's requirement of jam and jelly.

Since almost all provisions are frozen there are usually two large freezers on the farms with a combined capacity of about 500 litres.

Since most farms no longer have any cows—in Hagstedt nearly all the farmers have specialized in pig-fattening—all dairy products have to be bought. Bread too, together with a wide range of other foodstuffs, including international dishes like spaghetti, paprika, pizza, etc, are bought at supermarkets, where people pay particular attention to the special offers and buy these in large quantities for storing. Even at first glance it is clear that the movement from self-sufficiency to a mixed supply of food has brought with it a change in eating habits. Instead of the rather monotonous 'plain cooking' of around 1920, there is now a great variety of fare all the year round with strong influences from outside the region.

Sunday and weekday meals are now barely distinguishable since a great variety of foods are eaten on all days.

PHASES OF INNOVATION IN PRESERVATION METHODS

The shift from self-sufficiency to the 'mixed food supply' mentioned above is directly dependent on the innovations in preservation facilities.

From about 1910 food was preserved in glass jars in Hagstedt, but only meat was preserved in this way because the jars were too expensive for foodstuffs which came lower down on the scale.

From around 1925 tin cans were used for preserving and because these were cheaper than jars they were also used to preserve some fruit. After a return to traditional preserving methods during the war, people in the 1950s increasingly preserved food in jars, which had by then become more affordable. They also had the advantage of not becoming shorter, unlike the tin cans the rims of which were cut away after each use. From the early 1960s, households rented freezers at a large community freezer centre in the nearest town. Like the jars these were at first only used for preserving meat, that is, the most 'valuable' foodstuffs was given the most modern form of preservation. Around 1965 households obtained their own domestic freezers, the capacity of which was constantly enlarged, so that today every household has two large freezers.

The education of women played a decisive part in the introduction of the new preserving methods. All the women questioned were specially trained in rural domestic economy. The older ones had attended boarding school, the younger ones the rural domestic economy school and had attended further practical education courses. Cookery books were not important in the acceptance of the new methods. The technique of freezing was propagated by means of what were called 'educational afternoons' organized by the energy suppliers.

How far this educational campaign helped to break down the prejudices towards these new methods of preserving is made clear simply from the capacity of the freezers (about 500 litres). The technique of freezing is the most important method of preserving not only bought but also home produced food.Other techniques were thereby considerably restricted. Bottling was now used only for some fruit. Salting, clamping and drying were now hardly practised at all. So Gertrud Herrig's observations on the West Eifel apply here too: 'The tendency to give these up is especially apparent in the younger farming families or where the young daughters have translated the modern ideas from the agricultural schools into action.'[2]

FACTORS WHICH FAVOURED THE CHANGE IN STORING FOOD

I have mentioned only the improved food-preserving facilities in the domestic sphere, and these, together with the possibilities available in the food-preserving industry, provided the basis for changes not only in eating habits but also in the housewife's daily routine. But as in all processes of cultural transformation the combination of factors which actually brought about the changes were very complex 'for not every technical innovation immediately leads to the rejection of what has been customary up to then; this only happens when other factors are involved'.[3] Of these factors which are part of the whole development towards the consumer society, only a few can be indicated here. For example, the infrastructure as a whole, i.e., the improved transport facilities both in the industrial and the domestic sphere, has affected the storing of provisions. The mobility of the younger women is ensured by their having a driving licence and car, enabling them to drive to the nearest supermarket at any time. There are also dealers selling freezer food, for example, who come to the individual farms. So it is possible to do as much shopping as you want at all times. Other factors are connected with the economic changes in agriculture. Because the farms all specialize on a single line of production, products which used to be always available on the farms, such as a variety of meats, now have to be bought.

Around 1920 everything required in the way of food, apart from a few groceries such as spices, coffee, rice etc, was produced at home. Money was only spent on goods with a certain luxury character. At the same time self-sufficiency was not merely a necessity but an ideal. The amount of work which had to be done by the housewife to preserve all the food was not regarded in the same way as it would be today. Both the housewife's work and the food did not count as measurable values in themselves. Around 1920 self-sufficiency was the highest ideal in the thrifty mentality of the farmers in Hagstedt. The restriction of the farms to a single branch of production meant that the buying of food became increasingly unavoidable, and in the late 1960s and early 1970s there was a break-down of the inhibitions about having to spend money on everyday edible goods. At first the only food people bought, besides the usual groceries, was what they had previously produced themselves on their own farm, such as milk, butter, eggs, etc.But contact with the multiplicity of goods in the supermarkets increased their readiness to buy other things as well, to such an extent that in recent times people even buy fruit and vegetables, which a few years ago they still grew themselves.

HOUSEWORK AS MEASURABLE VALUE
The possibility of freeing themselves from self-sufficiency affected the women's attitude to work. The younger the women are, the more they see a measurable value in their own work and pay attention to the financial effectiveness of what they do, so the work on the farm is often more productive. The labour expended on planting, tending and harvesting the garden fruits, and then preserving them, appears less profitable than cheap special offers. Self-sufficiency is no longer worth the effort.

The older women regarded their housework primarily as a duty, and they hardly ever questioned the point of it. Demands for free time and convenience did not conform to the housewifely ideal. The social control, exerted for example by the neighbours, prevented them breaking out from the traditional values of the village. There was no room in the housewife's economic thinking for perceiving the relationship between the labour expended and the result achieved. The following argument of one of the younger women would never have been accepted: 'We don't have raspberries and strawberries in the garden any more, now that I don't dig it myself; my husband ploughs it up and disturbs the fruit.'

WHAT IS DEMANDED OF FOOD

Nowadays feelings of duty, necessity and thrift are scarcely the motivations which make people try to be self-sufficient. If fruit and vegetables are grown and preserved at home in large quantities, it is for different reasons. The decisive factor today is the recognition that produce from one's own garden is healthier and more tasty than the frequently sprayed vegetables from the greenhouses, or the products of the food industry, which are treated with all kinds of preservatives. The demand for deliberately healthy eating is slowly beginning to encourage a revitalization of the declining self-sufficiency in Hagstedt too. But in Hagstedt the awareness of the danger of damaging one's health through bad eating is not great enough for any radical consequences of this development to be apparent. In view of the revitalizing tendencies in the surrounding towns, which have advanced further than in Hagstedt, the older Hagstedt women who have held most strongly to home-produced food seem to be 'completely up-to-date with current trends'.

It seemed to me that for most of the women I interviewed the demands for a healthy diet were not nearly as important as the demand to eat food that was tasty and had variety. This led to continually increasing demands. In 1920 it was still impossible to object to the daily vegetable stew, whereas today it is very rare for the same meal to appear on the table for several days running. In 1920 it was only permitted for family members to express special wishes on dietary matters under particular circumstances. People only paid attention to personal needs if these were based on age or illness. Out of consideration for the health of the elderly, stocks of dried prunes and parsley root were kept, for example, to help their digestion. Mrs T. said: 'Our granny was terribly fat, she couldn't get up, so she had to do a lot for her digestion. We always had two large baskets of dried prunes in store for her.' It was also said that 'a chicken is killed only if the chicken is ill, or the farmer.'

The more self-sufficiency is abandoned, the more weight is attached to the wishes of individual family members. Above all much more notice is taken of the children. In some families not only is special food (e.g., sweets) bought for them, but also the garden is used for growing only what they particularly like eating.

The families have particular fashions in eating, which follow each other in rapid succession as people get fed up or sick and tired of them. The fatty foods which were popular in the past are hardly eaten at all today, and groats, once the breakfast food, have had to take a much lower place in the functional hierarchy; in one instance they were now only fed to the dog.

SUMMARY

The change in rural food storing in Hagstedt is not determined merely by technological advances. The possibilities opened up by the arrival of the preserving jar, the tin can, and the deep-freeze only brought about real changes when many other factors were also involved. As Hörandner says: 'Research on the "Preservation of Meat, Bacon, and Sausages in Pre-Industrial Times in Europe" has shown that preservation by salting, drying, curing (smoking), and laying in fat (or oil) was dominant in the countryside until after the Second World War.'[4] A fundamental change in the significance of storing food did not come about until the late 1960s with the emergence of extensive specialization in agriculture, improved infrastructure, with the possibilities of shopping, and other factors. Only then did the farms move further and further away from the principle of self-sufficiency, and since then the storing of provisions is no longer to be equated with self-sufficiency. Today's store of food mostly consists of large quantities of provisions bought from the supermarkets. In the past preserving methods determined eating habits; today it is these habits that determine the storing of food.

NOTES

1. Hörandner, Edith, 'Storing and preserving meat in Europe: a historical survey', *Food in Change* (ed) A. Fenton and E. Kisban, 1986, p 53.
2. Herrig, Gertrud, *Ländliche Nahrung im Strukturwandel des 20. Jahrhunderts*, Meisenheim am Glan, 1974, p 190.
3. Herrig, p 194.
4. Hörandner, p 57.

RELIGIOUS SIGNIFICANCE OF FOOD PRESERVATION IN INDIA: Milk Products in Hinduism

Mahadev L. Apte and Judit Katona-Apte

The institution of the sacred cow has played a significant role in the historical development of the Hindu religion. Milk and its various processed and preserved forms have achieved ritual status. A general framework of the preservation processes is proposed; and the question how and why the products are significant is discussed.

Food preservation techniques have been known to human beings since antiquity. The early hunter-gatherers were probably familiar with the process of drying meat and vegetable products to preserve them for later consumption. The life-style of early humans was such, however, that they did not need to preserve their foods on an extensive basis. It is only with the advent of animal husbandry and agriculture that food preservation became important. The nomadic pastoralist societies developed preservation techniques in order to preserve the milk of their domesticated animals. This was particularly relevant because of the varying ecological conditions that were met by the pastoralists as they travelled through different territories during the seasonal cycles.

Though it evolved to be an advanced civilization, the society and culture of India have preserved their pastoralist traditions in the institution of the sacred cow. While the cattle complex was the most crucial aspect of the nomadic Aryans who entered India from the Northwest, the indigenous Dravidian civilization and other tribal populations were equally conversant with animal husbandry, as indicated by the various Indus valley seals that depict bull figures (Basham 1959). The elaborately developed dairy rituals among the Todas of the Nilgiri mountains (Rivers 1906) are indicative that dairy herding is an old tradition in India.

It is not surprising that cattle have played a significant role throughout India's five thousand years of history. While numerous food preservation techniques have been developed in different regions over a period of time, our concern in this paper is primarily with the ritual and symbolic significance of various dairy products and the role played by the food preservation processes involved in it.

We will start by offering a general framework of preservation processes; then will concentrate on the discussion of how and why milk and its various preserved forms are important within the context of Hindu rituals and religious beliefs. Our approach will be both diachronic and synchronic. We will first present the reasons for the significance of dairy products in Hindu rituals; then the historical development of the sacred cow cult will be stated within the context of which many of these rituals will make sense.

There has been much discussion of the institution of the sacred cow in anthropological literature regarding its ecological, agricultural and economic functions (Anderson 1977; Harris 1965, 1966, 1974, 1978; Heston 1971; Simoons 1961). The assimilation of these aspects in religious ideology and its consequences are reflected in current Hindu cultural thought and in the Indian political party system (Simoons 1973).

FOOD PROCESSING: A GENERAL FRAMEWORK
A key to food preservation (a form of food security) is food processing. In our view, therefore, it is preferable to think in terms of food processing as a more basic concept related to food habits in general. Preservation is defined here as a processing technique that keeps food from spoiling through the action of some undesirable organism (e.g. bacteria, insect, etc). In functional terms, food can be processed either for immediate use or for preservation. Many activities such as cleaning and peeling, for instance, are undertaken for purposes of either immediate consumption or preservation.

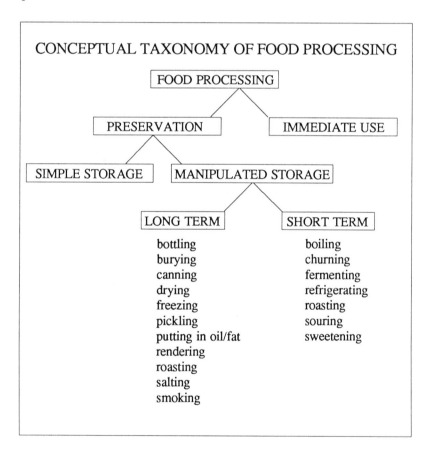

CONCEPTUAL TAXONOMY OF FOOD PROCESSING

FOOD PROCESSING

PRESERVATION — IMMEDIATE USE

SIMPLE STORAGE — MANIPULATED STORAGE

LONG TERM — SHORT TERM

LONG TERM	SHORT TERM
bottling	boiling
burying	churning
canning	fermenting
drying	refrigerating
freezing	roasting
pickling	souring
putting in oil/fat	sweetening
rendering	
roasting	
salting	
smoking	

Figure 1 is a depiction of a conceptual taxonomy of food processing; the list presented is not exhaustive. Preservation is a form of storage which can be either simple or manipulated. In simple storage, food items are placed in a variety of containers or specifically designated areas of the living compound. Manipulated storage can be further classified on the basis of time element, i.e. how long does the food item need to be preserved? We can conceive of long-term versus short-term storage, although these are relative concepts and the division between them is somewhat arbitrary and often contextually determined.

In general, long-term storage can be said to be connected to the seasonal food production cycle and therefore to the abundance of particular foods at certain times of the year and their unavailability or scarcity at others. When the agricultural cycle is closely linked to the seasonal cycle and to water availability, societies go through stages of plenty and scarcity. In cultures around the world many major festive and religious holidays are linked to the harvest time when the staple grains are plentiful. In India, for example, the most important festivals, *Divālī* (festival of lights) and *Pongal* occur right after harvest in October or early November.

The most common processing techniques for long-term preservation are canning, coating with oil, drying, pickling, roasting, salting, smoking, and so forth. For short-term preservation, boiling, churning, fermenting, roasting, souring, sweetening and, in modern times, refrigerating are commonly used (see figure 1). A variety of dishes for everyday consumption and for special festive occasions, such as religious holidays, are prepared by using some of the processes involved in short-term preservation.

There are folk beliefs associated with food not properly preserved in terms of the presumed consequences of consuming it; conversely, some preserved food items are believed to have medicinal or healing properties. Popular cookbooks in Maharashtra, for instance, list traditional techniques developed for food preservation and also suggest the medicinal properties of many foods thus preserved (Ogle 1981; Wagle 1964).

It is the primary thesis of this paper that the relationship between religion, i.e. Hinduism, and products of the cow is closely associated with the processes necessary for transforming milk, the basic product, into other products which are preserved forms of milk itself. This thesis is elaborated below. It is necessary, however, to first discuss the significance of cow products in the Hindu ritual system.

SIGNIFICANCE OF MILK PRODUCTS IN HINDUISM
All cow products are crucial in Hindu rituals because of their functional role in the most fundamental ritual dichotomy of purity/pollution. This dichotomy pervades all aspects of Hinduism including behavior, action, social ranking, and cultural ideology in a variety of social contexts. According to one of the theories on the origins of the Hindu caste system, the purity/pollution dichotomy is at the basis of cast hierarchy (Dumont 1970). Notions of purity and pollution influence personal interaction

within the context of household, family, circle of friends and acquaintances, and community. Rules of purity/pollution are most crucial in matters of cooking and eating. Hence the significance of products of the cow as purifiers.

Basically, the idea of purity and pollution is closely associated with the concept of eternal salvation (*moksa*), in which one's soul (*ātmā*) merges with the quintessential metaphysical force (*brahman*). In order to achieve this, human beings have to remove themselves as much as possible from all biological processes that perpetuate the cycle of life and death. All bodily functions and processes, and substances discarded as a result, are thus considered polluting. Defecation, urination, spitting, menstruation, removal of body hair or nails, and so forth are all polluting acts. Similarly, processes of birth and death are polluting. In any religious activity involving worship of gods one must be in as ritually a pure state as possible in order to acquire merit. This means that one has to go through purification processes, as it is not possible to avoid being polluted due to the necessity of engaging in bodily functions.

The purifying agents in the conceptual and ritual system are water, fire, and sacred substances such as cow products, including milk[1] and its derivatives, through the application of various food preservation techniques. The five products of the cow (*pañcagavya*) (milk, curd, butter, urine and dung) have great purifying potency especially when combined in a single mixture (Basham 1959: 319). Such a mixture is consumed during special ceremonies such as ancestral rites. The processes of purification may be symbolic or extensive depending on the degree of pollution and on how orthodox one is. An orthodox person's entire sphere of daily activities can be dominated by such concerns. Hindu scriptures applied this dichotomy so thoroughly as to classify all substances, especially food, into various hierarchies, using the criteria of purity and pollution.

Cooking has to be carried out in the most pure state, since food is very vulnerable to pollution and yet has to be offered to gods as *naivedya* ('food offering to god[s]') in a pure state (Apte and Katona-Apte 1981). After the worship *naivedya* becomes *prasād* ('consecrated food') which is consumed by devotees. The very act of eating is polluting, since it involves the process of ingestion and the use of a bodily substance such as saliva.

Within this framework of purity and pollution, it is understandable why purifying agents are necessary and why only purified foods are offered to gods as propitiation in worship. In the hierarchy of purifying agents, cow products are ranked the highest due to their close association with gods. Ordinary human beings bathe with water to transform them from the polluted to the pure stage, for example, but in many Hindu temples gods are bathed in milk which is purer than water. One of the most sacred activities that is undertaken in many temples is to bathe the deities with *pañcamṛt* ('five types of nectar'), a mixture including milk, curds, clarified butter, honey and sugar. The symbolic significance of the inclusion of three dairy products, two produced by food processing, is noteworthy.

There are other aspects of Hinduism that emphasize the importance of milk and its products. Milk, curd, and ghee have been important items in the diet since ancient times (Basham 1959: 194). Brahmins, who are more ritualistic than others, use more milk (Katona-Apte 1977). Persons of higher caste will not accept water-cooked food from persons of lower caste, but will accept food fried in ghee. Most of the foods offered to or associated with deities are milk, or a preserved form of it (for an elaboration of what foods are offered to which gods, see Apte and Katona-Apte 1981). In Maharashtra, a millet bread called *bhākrī* when prepared with milk instead of water can be consumed even by persons in a 'pure' state who cannot consume foods made with water, and by those who are 'fasting' and thus should not consume foods made of millet (Katona-Apte 1976).

USE OF MILK PRODUCTS IN HINDUISM
When considered in this light, it seems reasonable to link the various stages of preservation of milk which transforms it into other products with differing levels of effectiveness as purifiers. It is generally the practice in most Hindu households that, as soon as milk is delivered, it is boiled, and this is repeated several times a day, as a means of preservation. The cream that accumulates at the top is then removed and preserved for later use with a souring agent. Whatever milk is left over at the end of the day is converted to yogurt by the addition of just a little portion of left-over old yogurt (souring by culture) for consumption the next day. As there is usually not enough yogurt for consumption, it is converted into buttermilk by adding water and churning.

The cream that has been preserved with the aid of a souring agent is churned to produce butter and buttermilk. This butter can be consumed as it is (e.g. with unleavened bread), or is saved for several days before being converted still further by heating it over a low fire into its most preserved form, clarified butter or 'ghee' as it is popularly known all over India. As this product is pure fat, containing no water or solids, it lasts without spoiling for an indefinite period of time. Once transformed in this fashion, this milk derivative has a variety of uses, such as being consumed with rice, sprinkled over bread, or as an ingredient of sweets. It should be remembered that the amount of ghee obtained from a given quantity of milk is very small, thus a plentiful ghee supply along with its extensive use in the preparation of food items is considered a mark of riches.

Thus the milk preservation progression is as follows: milk to boiled milk to yogurt to butter to ghee. In terms of purity, buttermilk is the lowest as it is a mixture of milk and water, yet it is still purer than water. Note the comment of Abbé Dubois who wrote in the first half of the 19th century (Dubois 1906: 187):

Curdled milk diluted with water is a very favourite drink with Hindus. It is usually Sudras[2] who prepare and sell this refreshing beverage. Although, generally speaking, there is more water than milk in the mixture, Brahmins have no scruples in partaking of it, and if anyone reproaches them with thus using water drawn and handled by Sudra, they reply that the curdled milk, which has come from the body of a cow, cleanses it from all impurities.

Ghee is the most highly preserved form of milk and it is the purifying agent of the highest status. When offering food to the gods, for example, ghee is the final item added to purify it. In certain regions of India, the most important task of purification at all festive occasions, when food is served, is done with the addition of ghee to the prepared foods just before guests start eating (Apte and Katona-Apte 1975).

Historically, ghee achieved such a significance because of its extensive use in the most important religious ritual of the Aryans called *yadnya* (also spelled *yajna*) 'sacrifice'. Descriptions of these sacrifices as performed by Brahman priests for their kings and noblemen are found in the Vedic and Epic literature dating back to 1000 BC and the most important offering appears to be clarified butter (Rawlinson 1952: 31). Butter and ghee are mentioned in the 'Hymn of the Primeval Man' of the *Atharva Veda* which describes the first cosmic sacrifice (Basham 1959: 240).

The extent of the sacrifice reflected the social status of the host and often continued for several days with guests who had come long distances to attend. Both Hindu epics, the *Rāmāyana* and the *Mahābhārata*, describe the performance of such sacrifices. As ghee, the ultimate preserved milk product from the cow, acquired such an important role in these rituals, it became the most sacred symbol of Hindu religious ritual.[3]

In many temples in present day India the foods offered to gods become consecrated and are then distributed to devotees. Often such foods consist of sweets prepared with ghee. The use of ghee in these foods is also important because they need to be preserved during travel. Individuals who go on a pilgrimage to specific temples generally offer money to the priests, who then give them consecrated food prepared with ghee. Devotees eat some of it, but carry the rest home, after travelling several days to distribute it among their relatives. Religious Hindus believe that eating such consecrated food results in the accumulation of religious merit towards the final salvation from the cycle of life and death. The fact that such food is fried in ghee is a major factor in its preservation over a period of time. As Breckenridge (1986: 41) points out, divine foods such as sweets are both 'preservable and portable ... they can be consumed subsequent to the ritual context in which they were generated, and they could be redistributed to persons not attending the worship ceremony'.

One of the most common forms of worship (*pujā*) of Hindus all over India is the *Satyanārāyana Pujā* which is generally performed to give thanks for the successful completion of a major activity (e.g. wedding, building a house, concluding a business transaction, recovery from illness, and so forth). The consecrated food associated with this worship which is offered as *prasād* to all guests is made by frying farina in large quantities of ghee (with sugar and various kinds of spices and nuts added). The quality of such an offering is usually determined by the amount of ghee that is readily noticeable.

SUMMARY AND CONCLUSION

The institution of the sacred cow has played a significant role in the historical development of the Hindu religion. As a result, milk, a product of the cow, and its various processed and preserved forms have achieved ritual status because of their functions as purificatory agents in Hindu rituals. Among these products, ghee, the most highly-processed and preserved item, has the highest status as a purificatory agent in religious acts of worship, consecration and feasting. This is, therefore, a prime example of the symbolic significance of food preservation in human sociocultural existence.

NOTES

1. This also includes buffalo milk and its derivatives. No other product of the buffalo, however, is considered sacred.
2. Sudras are members of the lowest hereditary class in the Hindu social hierarchy.
3. That this situation is not unique to Hinduism may be demonstrated by the fact that in Christianity and Judaism the food item most commonly used for ceremonial purposes is wine, a preserved form of grape.

BIBLIOGRAPHY

Anderson, E.N. , Jr. , 1977. 'On an early interpretation of India's sacred cattle', in *Current Anthropology*, 18(3): 552.

Apte, Mahadev L. and Judit Katona-Apte, 1975. 'The left and right side of the banana leaf: Ethnography of food arrangement in India', paper presented at the American Anthropological Association annual meeting, San Francisco.

—1981. 'The significance of food in religious ideology and ritual behavior in Marathi myths', in A. Fenton and T.M. Owen (ed), *Food in Perspective,* Edinburgh (UK), John Donald Publishers, pp 9-22.

Basham, A.L., 1959. *The Wonder that was India,* New York, Grove Press.

Breckenridge, Carol A., 1986. 'Food, politics, and pilgrimage in South India, 1350-1650 A.D.', in R.S. Khare and M.S.A. Rao (ed),*Food, Society and Culture,* Durham, N.C., Carolina Academic Press, pp 21-53.

Dubois, Abbé J.A., 1906. *Hindu Manners, Customs and Ceremonies,* 3rd edn, Oxford, Clarendon Press.

Dumont, Louis, 1970. *Homo Hierarchicus: The Caste System and its Implications,* Chicago, University of Chicago Press, pp 42 ff.

Harris, M., 1965. 'The myth of the sacred cow', in A. Leeds and A.P. Vayda (ed), *Man, Culture, Animals,* Publication no 78, American Association for the Advancement of Science, Washington, DC, pp 217-28.

—1966. 'The cultural ecology of India's sacred cow', in*Current Anthropology* 7: pp 51-60.

—1974. *Cows, Pigs, Wars and Witches. The Riddles of Culture,* New York, Random House.

—1978. 'India's sacred cow', in *Human Nature* 1(2): 28-36.

Heston, A., 1971. 'An approach to the sacred cow of India', in *Current Anthropology,* 12: 191-209.

Katona-Apte, Judit, 1976. 'Dietary aspects of acculturation: Meals, feasts and fasts in a minority community in South Asia', in M. Arnott (ed), *Gastronomy: The Anthropology of Food Habits,* The Hague, Mouton, pp 315-26.

—1977. 'The socio-cultural aspects of food avoidance in a low-income population in Tamilnad, South India', in *Journal of Tropical Pediatrics and Environmental Child Health,* April, pp 83-90.

Ogle, K., 1981. *Rucirā,* Poona, Kirloskar Press.

Rawlinson, H.G., 1952. *India: A short cultural history,* 2nd revised edn, New York, Praeger.

Rivers, W.H.R., 1906. *The Todas,* London, Macmillan.

Simoons, F.J., 1961. *Eat Not This Flesh,* Madison, Wisconsin, The University of Wisconsin Press.

—1973. 'The sacred cow and the constitution of India', in *Ecology of Food and Nutrition* 2(4): 281-295.

Wagle, S.G., 1964. *Rucivaibhav,* Bombay, B.L. Pathak Prakashan.

DOMESTIC STOCKPILING BETWEEN 1920 AND 1984 - COMPARISON OF URBAN AND RURAL CONDITIONS IN NORTH-WEST GERMANY

Anke-Sabine Barth

Are stores of food kept at home only under particular circumstances, or is stockpiling an unwritten law for every household? This was the question I asked myself after my investigation of a rural community in North-West Germany. The twelve interviews I had conducted had all been with farming households, and a corresponding investigation of urban households promised to provide an interesting contrast. The area of investigation chosen was Osnabrück, the nearby district town. It has 150,000 inhabitants and is the third largest town in the Land of Lower Saxony, so it was clear from the outset that my results could not be representative. However, my 1983 investigation of the rural communities had already emphasized that as far as the storing of provisions was concerned, there were few deviations from generally widespread practice. Hence the four interviews conducted in Osnabrück may at least form the basis for some hypotheses regarding the question posed at the start of this paper.

The investigation was undertaken in 1984. The interviewees were female, between 31 and 78 years of age and belonged to the bourgeois middle class. They spoke about their own household management during the time after they were married. Supplementary details from later years were also recorded if there was an increase in food-storage then. In this way the period between 1920 and 1984 was fairly well covered.

The questions dealt not only with the preserving and storing of foodstuffs but also with the interviewees' training in household management, their own housekeeping, and their professional activities outside the home.

This last point proved to be of particular interest as three of the four interviewees had jobs which occupied them for half the day, or even the whole day. They therefore had little time left for running a household.

Were they able nevertheless to keep a store of provisions?

The oldest interviewee was married in 1926. She had a full-time job as well as looking after a household of four. She said: No, at that time she had not kept a store of meat. There were no freezers then. Everything was bought in small quantities from the butcher whose shop was only two houses away. The meat products were kept for one or at most two days in the cellar or larder. She had also bought dairy products and bread in small quantities as required. The storing of food was thus transferred to the shops in the neighbourhood.

The only goods bought in large quantities, as a seasonal addition to the shopping list, were vegetables and fruit, which were preserved and stored. The most important reason for doing this was the special flavour of home-salted sauerkraut and home-bottled fruits, since stored fruit and vegetables could still be bought in the market in winter for slightly higher prices.

This rudimentary form of food storage was practised by the interviewee because of the demands of her work. The food supply situation in towns made it possible for her to have a job as well as running a household. In the years of shortage during and after the Second World War she again made use of what she had learned in her training in domestic economy.

The second oldest interviewee spoke about the time after 1951. She had studied agriculture and ran an urban household of seven, which owned a small farm.

In the first years of her marriage slaughtering was still done on this farm and she carried out all the necessary preserving processes herself. Later, however - mainly because of the considerable saving in time and effort - meat and sausage were bought at the nearby butcher's. The electric refrigerator was also introduced into the household and provided improved storing and cooling facilities. It could also be used for storing other foodstuffs, such as dairy products and vegetables.

Fruit and vegetables, however, were an exception in this household too. The harvest from their own garden was always eaten fresh, but she bought quite large quantities of fruit and vegetables at the market or direct from farmers, and these were preserved - again mainly for the sake of their particular flavour.

The third interviewee had received no training in household management, but according to her own account this was not the reason why she did not store food. As a housewife she had compared prices and realised that it was just as cheap to buy most food in the shops as it was to bottle or store it oneself. When her two children were older and she took a job as a night sister at a hospital, she found she had to buy food in advance. Since then (1969) she had used the storage facilities provided by a deep-freeze.

This interviewee too said that she nevertheless liked to preserve fruit and vegetables, although she had to buy them.

This example makes clear on the one hand the decisive importance of the genuine need to lay in a store of provisions, and on the other hand how little attention is paid to the economic aspect of the particular products to be preserved. The preserving of such produce as fruit and vegetables is apparently a sophisticated relic of the practice of storing provisions at home.

The youngest interviewee had attended a housekeeping school before her professional training. She spoke about the period after 1977. Besides running her four-person household she also had a part-time job.

From the very beginning her housekeeping was based on the storage facilities provided by the refrigerator and deep-freeze, since she lived a little outside the town centre and so had a longer way to go to the shops than the other interviewees. She visited the shops at regular intervals, buying in bulk and storing a certain amount of vegetables, fruit and meat in the deep-freeze. She also made small purchases of dairy products, sausage and small quantities of fresh fruit and vegetables for the day's requirements.

In contrast to the three older interviewees she made no mention of bottling fruit and vegetables. In her household the deep-freeze had completely taken over these functions, since the family liked bought fruit conserves and frozen fruit - as well as home-frozen, of course.

The examples given do not of course provide a representative picture of food-storing in Osnabrück households in the last sixty years, but they do highlight the differences between urban and rural households in the same period. This is shown by a comparison with the village of Hagen in the Teutoburger Wald ten kilometers south of Osnabrück. It lies in a valley in the Teutoburger Wald and, despite its economic connections with the town, it still retains its predominantly rural character.

In the period around 1926 most of the food required by the urban household was provided by daily visits to the shops, whereas the rural farm at that time was still almost completely self-sufficient. For one thing people were in a position to produce all their own food, and for another they made too little money to be able to buy food from shops for the household. Their pride as farmers made them regard the purchase of food as shameful: people who had to 'buy extra' were obviously no longer able to produce sufficient yields on their farm. This attitude is surely also a reason why it was only with difficulty that food shops became established in purely farming districts.

Around 1950 changes in the structure of agriculture also affected the domestic storing of food. Technical innovations were introduced into housekeeping as into farming as a whole. The significantly longer storage life of bottled vegetables and fruit, and not least the improvement in flavour in comparison with the dried varieties, changed the country-women's attitude from initial wariness to a delight in innovation.

The deep-freeze, for example, was in use on some farms in the 1950s, but only appeared around 1970 in the urban households I examined. To understand the impact of the revolution in preservation technology represented by the deep-freeze on rural food storage, it is necessary to realize how vitally important food provisions were for the farming family, and how much time and effort were spent on the preserving of meat, fruit and vegetables. One of the farmer's wives interviewed described her first impression of the new freezing equipment: 'It was wonderful! Everything could be put straight in without any big preparation!'

Urban households, on the other hand, with no livestock or gardens, had no foodstuffs that needed to be preserved, so the deep-freeze was not nearly as necessary. It was only gradually that the urban housewife adapted herself to a changed food supply with the many varieties of deep-frozen dishes.

Although the technique of freezing made preserving food much easier, the intensive storing of food in the countryside nevertheless gradually declined. The main reasons for this are to be found in the increasing specialization on farms. On the one hand it was no longer possible to produce everything one needed oneself, on the other hand people now made enough money from selling their produce to buy food from the shops.

Since the 1970s the main difference between a farming household and an urban one has been that one still has access to the basis of its previous self-sufficiency, whereas the other has depended entirely on the goods available at the market for its food supply. The opposite extremes represented by the 'purely farming household' and a 'purely urban household' are gradually converging, which shows that the intensity of food-storage depends on particular basic circumstances. Training in household management plays hardly any part in this. What is important are the available sources for obtaining food. If people are restricted to their own produce they have to lay in an extensive store of food, but if they are also financially and geographically able to go to the shops, they will take advantage of this and their self-sufficiency will correspondingly diminish. This means that there is no stockpiling of food if there are no opportunities for self-sufficiency or if the opportunities for shopping are available instead.

The remnants of a few particularly favoured methods of preserving are retained: home-made jam, bottled fruit and pickled vegetables (e.g. pickled cucumber) are still popular. Under the influence of the ecological movement in particular 'home-made' food has become not merely a hobby but something like a prestige object—both on the farm and in the urban household.

Prestige object and relic of a by-gone age of intensive food-storing? Everything seems to point to this conclusion, but the storing of provisions is not a phenomenon that will disappear from history, or from cultural history. The developmental trends described above presuppose that the general state of the food supply remains constant and its infrastructure becomes increasingly refined. An economic crisis or any other disruptive factors affecting this infrastructure would entail an immediate return to the stockpiling of food. Stockpiling should be regarded not only as a phenomenon subject to political circumstances,but also as a visible indicator of the current conditions in town and countryside.

FOOD AS A MEDIUM FOR PRESERVING CULTURE

Beatriz Borda

This paper approaches its subject through empirical studies of the behaviour of Latin American women in Scandinavia, particularly in connection with the cooking of their traditional dishes.

The subject of the present conference is conservation. Webster's Dictionary *defines conservation as 'conserving, protection from loss, waste, etc; preservation' and 'the official care and protection of natural resources', while the verb conserve is defined as 'to keep from being damaged, lost, or wasted; save'. These many senses give us the possibility of looking at conservation in the light of quite different perspectives.*

My focus here will not be the different techniques of food preservation but food-ways as a means of preserving ethnic culture in an alien cultural environment.

MATERIAL, METHOD AND PURPOSE

In Sweden today there are about 13,000 Latin-American refugees, of whom some 1,800 live in Malmö. Most of them belong to the middle class and are well educated, often with university degrees. There are also a fair number of working-class refugees, and a smaller number of people from over-populated city slums. The reason that most of them fled to Sweden is political—a left-wing protest against the ruling regimes. Political murder, torture, and systematic persecution of all opposition became an everyday experience, at the same time as the material circumstances for the majority of the population saw a dramatic deterioration. Refugees began to pour across the borders. Many were given asylum in Latin-American countries further north, while others were admitted to various countries in Europe, among them Sweden.

There are differences in the culture and life-patterns of the different nations in Latin America and between the different classes within each country. They nevertheless share so many characteristics that it is possible to talk of one Latin-American culture as distinct from Swedish culture. The purpose of this study is to investigate the role played by meals and their composition in expressing the social relations and ethnic identity of the Latin-American refugees.

The empirical evidence has been assembled through participatory observation and interviews with 35 Latin-American families in Malmö.

KEEPING A CULTURE ALIVE

Why do people eat what they eat? This question is fundamental to ethnological research on food-ways and eating habits and it has become clear that, particularly amongst people who meet foreign cultural patterns in daily life, food and food-ways are of great importance.

It has often been contended that food preferences depend on what is accessible in the environment or what is possible in the technological and economic circumstances. My own studies of ethnic food habits among Latin-American immigrants in Sweden show that the way we eat depends as much on what culture permits as on what culture insists is appropriate. In the context of my topic, ethnic foods-ways, the concept of culture as a tool for identity-maintenance and classification of the social world raises many interesting questions. What structures can we find in the process of ethnic revitalization? More particularly, how do specific individuals —clearly agents in the process of revitalization—contribute to this process? Who is authorized to make judgments on tradition and declare that 'this' is genuine but 'that' is not, 'this' is worth being saved and ritualized, but 'that' is not (Arnstberg 1985, Gerholm 1985)?

The place of food in everyday life is both taken for granted and imbued with cultural, social, and personal significance. Everyone must eat, usually several times a day, and this situation leads to highly patterned and regularized behaviour. Indeed, most of human life is similarly patterned and repetitive, otherwise everyday life would be lost in details. But the biological need for food and the social act of eating combine to give food patterns a particular meaning, a kind of cultural power. This can be seen clearly among ethnic minority communities.

TRADITION AND IMPROVISATION IN EVERYDAY LIFE

Many Latin-American families residing in Sweden treat food and eating as central concerns which require much attention and discussion; this is their means of creating a way of life of their own, keeping a great deal of their common cultural heritage. Food preparation is also a subject of endless elaboration, depending on the degree of commitment among members of the same ethnic group, making special arrangements of typical food distinct from other situations of food preparation, service, and consumption. The amount of time and work spent in preparing this kind of food is repaid by the emotional satisfaction often engendered by sharing not only food but also a common adventure.

A focus on food will provide insights about how and why ethnic groups maintain, develop, or change certain patterns of food-related behaviour when living in a strange host society.

The rhythm of life in Sweden makes it difficult for the families to retain the old pattern of mealtimes. Here each member of the family has to follow his or her own personal daily timetable. This means that a considerable part of the cooking which was once done in the home has been replaced by various catering services: schoolchildren have lunch at school, and if both husband and wife have jobs, no lunch at all is cooked during the working week. The evening meal has thus become the 'communion' at which all assemble once again.

Meals enjoyed together with people from outside the household can be termed 'social meals'. Sunday lunch as a social meal is common in Latin America in conjunction with visits by relatives and friends. The same pattern survives among the refugees in Sweden, who maintain frequent dining contacts and meet each weekend in each others' homes to share meals.

> When we have guests we usually treat them to Argentinian dishes which are difficult to cook in Sweden and which we miss, because they were part of the everyday cooking in our homeland. A starter which we call *antipasto*, to stimulate the appetite, followed by a main course with beef, and finally some sort of dessert, ice-cream, cake, or fruit salad.

Dishes consisting of ingredients that are hard to acquire, and which need much time and effort, are made only when Swedes are invited. This is a way to fulfil Swedish expectations of this exotic Latin-American culture. People are also keen to show that they have a culture of their own, and that it is different from that of the Swedes. This is confirmed by the 'typically Latin-American dishes', which many of the refugees only learned to cook after they came to Sweden, and which they like to serve to Swedish guests.

> When I serve my guests Chilean food, I try to give it 'the right taste', since a Chilean dish on the table is a taste, a smell, a picture of the homeland.

Weekend meals are generally more lavish and ample in both quantity and quality, but they do not differ from everyday food in Latin America. Food for festive occasions in Latin America shows greater variation and profusion. At such meals they serve several cold and hot courses, and a range of desserts and cakes (Norton 1970).

Daily rituals in the home country have new meaning in Sweden.

My own empirical studies show that there are three mechanisms operating in order to make a success of the sense of community through eating. There is the investment of large contributions of time, money, personal property, and work. There is a communion involving group activities and rituals of work, play and celebration. Finally, there is a sort of transcendence, or the surrender to a total involvement in a system larger than the individual members.

The difficulties in acquiring the ingredients required for the preparation of Latin-American food, either through special shops or through relatives residing in the homeland, transform ethnic dishes into special meals reserved for special occasions.

Another factor is time. Everyday life in a highly industrialized society like Sweden does not give the opportunity of spending too much time in the kitchen. Daily meals may take a short time on busy days, while the slow-cooking Latin-American food is a weekend project. A reasonable amount of time and effort is expended on a meal like this, since it has to be prepared once a week.

Many Latin-Americans have a few specialities or a particular style of cooking with which they are identified. Some of the women, for example, are well known for their chilli dishes, while some of the men specialize in steaks. Standard dishes such a spaghetti, meatballs, vegetable dishes, or macaroni are accepted for everyday meals when both time and inspiration are short.

HERITAGE AND CHANGE

Cooking in Latin America is regarded as typical women's work. The woman's role in preparing the food is of great importance. She is the one who buys, cooks, and serves the food. She is in other words responsible for the entire process from natural product to cultural product, and this is seen as her given duty. Although cooking is not viewed as a particularly prestigious activity, a woman's culinary skill means a great deal for her esteem.

Ethnic meals eaten together with people outside the immediate family are of great importance, especially to the women. In this stituation a woman's cooking style becomes an integral part of her personality in the eyes of the others, and culinary skill is part of being a true Latin-American. The opinion of the group is important. Thus, during the meal, comments are always made to let the cook know that the food is appreciated and tastes like 'typical' Latin-American cooking. Silence is feared because its message of disapproval is not to be mistaken: 'Not much can be said of the cook's cultural competence'.

Leftovers are not stored. Some foods can be left in pots on the stove for later meals during the day. But they are never frozen to save them for following days. The food is usually kept according to the principle that someone may eat it, and that, if no one does, it can just as easily be thrown away. Re-use or rearrangement of leftovers is considered a legitimate meal only if other items are added and if there is some evidence of effort and creativity in the new dish.

The arrangement of Latin-American dinners shows the level of commitment and the operation of a cultural mechanism, 'group communion'. Such bigger events go beyond individual contributions and necessitate co-operative cooking and arrangements to achieve success, besides which they involve the sharing of diverse cultural heritages and traditions, which often develop into new traditions when confronted with the Swedish cultural context.

The nature of the commitment felt by the members of the ethnic group is clearly seen in food and eating patterns. The organization has established standards which evolve into cultural patterns that become natural for Latin-American immigrants living in Sweden.

The culture of immigration generates special forms, exploiting tradition in distinctive ways and displaying a preoccupation with particular themes (Kirshenblatt-Gimblett 1978). For example, new combinations may appear as a result of disagreement over the proper way to make a special dish as prescribed by tradition.

Food as a microcosm of a shared lifestyle can be applied to any number of people interacting socially, from a family to the people working in a factory.

Making Latin-American food requires many ingredients and a number of steps. Because of the work required, festive meals are seldom prepared by only one person. As a result, Latin-American dinners nearly always call for collective effort and are characterized by a great deal of sociability. Friends and relatives gather together, sharing thoughts and feelings while participating in a communal activity ostensibly aimed at making food to be consumed later in a spirit of friendship and equality.

The women participating are usually not related, but they share in common the fact of being Latin-Americans in Sweden and having common cultural values. In addition, of course, they are involved in an activity they like - cooking their national food. During the work the topic of greatest interest to the women concerns the different styles of Latin-American cooking. In the discussions more is involved than the simple exchange of recipes. Various kinds of Latin-American food and different ways of cooking it are explained according to the women's ideas of the influence of geographical origins and family traditions.

While a spirit of co-operation prevails most of the time, disagreement can arise as to the proper way to prepare some dishes. In the discussion the individual proponents of the different methods try politely but firmly to defend their way as the best and genuine one. And, not surprisingly perhaps, the making of Latin-American food, like many other activities people engage in, is subject to aesthetic concerns and criticism. In other words, people can have lengthy discussions regarding what constitutes 'genuine' Latin-American food. They talk about the taste and smell of different dishes, the best way to season each dish and the reasons for this.

The Association of Latin-Americans in Sweden selling their native food at a festival in Malmö.

Whether the discussion is about the ingredients, the procedure of apportionment, or the manner of cooking, the women express a serious concern for the 'right' way to make Latin-American food. Many of them are preoccupied with finding the original way, having a great sense of pride in their work. These women have a number of impressions and perceptions in common; experience of the conflict that often accompanies the life of immigrants in multi-ethnic urban communities and language differences which make it difficult for them to communicate in Swedish. They tend, therefore, to be separated from all but their immediate family members and countrymen. As a result they share a sense of isolation and fear about their place in Swedish society.

From this point of view, Latin-American dinners take on a great deal of importance for them as a social event. By moving their work out of a private kitchen and into a public one, the women earn recognition and are appreciated by a larger group beyond the boundaries of their own families. Through the interaction of the participants, the cooking of Latin-American food may be seen as an ordered means of socializing people in certain specific ethnic values. This becomes apparent in relationships formed between the sexes and between different generations. The ability to make good Latin-American food is seen as a skill to be emulated. Those with the longest experience are often acknowledged to have this ability. And among these cultural performers there is usually someone with a greater amount of cultural competence who acts as a perfectionist. In any case, through their interactions, the indvidual participants increase and refine their knowledge and their ability to cook 'typical' dishes.

Obviously, these occasions may serve to instruct the younger generation in the methods of preparing the traditional food. But perhaps more importantly, they are a forum for expressing and reinforcing certain values attached to the cultural heritage of the group. Old rituals are adapted and new rituals created to deal with recurring and unique events in people's lives. An ethnic community and its culture are defined as much by what is rejected as by what is accepted, by what is discarded as by what is retained. In this light, 'persistence' and 'attrition' need not to be viewed as passive acts, but as 'active' decision-making that shapes cultural continuities and discontinuities.

Another important feature of Latin-American co-operative cooking in Sweden is its contribution to stronger female identity among the women. The women also find Latin-American traditional cooking an opportunity to present a unique talent, to demonstrate an exclusive ability. By its very nature, the event also becomes a forum for 'performances' of various kinds, creating opportunities for the participants to show their individual skills.

In many households, the creative energies of immigrant raconteurs are unleashed by the situation of multiculturalism, in which world views, cultures, and languages clash and undergo massive and rapid change. The protagonists of their stories are often tricksters, who are considered to be without culture because they are between cultures. What they know from the old country they cannot use, and what they need in order to succeed in Sweden, they do not yet know. These tales are generally preoccupied with culture shock, name changing, linguistic and cultural unintelligibility, and the eccentricities of both immigrant and Swedish characters (Kirshenblatt-Gimblett 1978).

These tales and jokes also indulge in ethnographic detail in ways that reveal a preoccupation with cultural competence, contact, and change, as well as with the gap between Latin-American and Swedish culture. These mutual events exhibit the sense of communion of the group which increases through shared knowledge and work in other aspects of life, not just in food-related activities. The efforts expended in planning Spanish lessons for the children, for example, and in organizing leisure activities together are very important in establishing a feeling of belonging to a group as well as generating ethnic pride. The group gains meaning as a whole, becoming more than the sum of its parts. That is why the necessity for contributions of time and thoughtfulness are understood to be a small price to pay for a sense of belonging and a warm social environment.

To honour a special person, special foods are required, which means that the group planning the meal often decides that the occasion calls for something out of the ordinary. The women assemble at someone's home in order to prepare the dinner. Many of them gather not only for birthdays but also at New Year and Christmas, and so forth. Ethnic meals are important not only as an occasion for the preparation of typical dishes essential to certain events, but also as social events in themselves (Borda 1987).

Importantly, also, these types of interactions contribute to the shared experiences which develop the sense of unity and community that give rise to traditional events. Each event and each contribution to it adds to the experience, the values, and the significance of ethnic food.

REFERENCES

Arnstberg, Karl-Olov, 1985, 'The Concept of Culture as a Core Symbol', in *Methodological Questions* no 2, Nordiska museet, Stockholm.

Borda, Beatriz, 1987, 'Kost och etnisk identitet. Matens roll som identitetsmarkör hos invandrade sydamerikaner', in *Mera än mat*, (ed) Anders Salomonsson, Stockholm.

Gerholm, Lena, 1985, 'Revitaliseringens fenomenologi', in *Nord-Nytt*, 25/1985, Viborg.

Kirshenblatt-Gimblett, Barbara, 1978, 'Culture Shock and Narrative Creativity', in *Folklore in the Modern World*, (ed) Richard M. Dorson, The Hague.

Norton, Leonard 1970, *Latinamerikanskt kok*, Stockholm.

THE CONSERVATION OF FOOD AS A STRATEGY IN PEASANT ECONOMY: an example from the valley, fjord and coastal regions of the county of Sogn og Fjordane in Norway

Bjørn Fjellheim

In the literature, little attention has been paid to the conservation of fish; much more has been written about the fisheries and the trading of fish. However, focussing on the total process of conservation provides a good means of unveiling important social and economic factors and situations.

SOGN OG FJORDANE

One way to regionalize peasant societies is to discover how access to food influenced the living conditions of people in various parts of a chosen area. An example of this is the following description, from the end of the 18th century, of Sunnfjord, the middle part of the county of Sogn og Fjordane (fig 1).

In this respect, a distinction must be made between the valley people, the fjord people and the coastal people. The valley people do not use much fish and herring. Bringing it on horseback is both strenuous and costly, so that they must be satisfied with a smaller share of this. In its place, they must see to it that they have the meat and more abundant dairy products from cattle to resort to. This is why, generally speaking, one can say that the people high up in the valleys live better than the fjord people. The latter depend far too much upon their beloved herring, and therefore, when it fails, know no solution. Concerning the coastal people, in years when fishing is bad, fish may be their only support and it often substitutes for both meat and bread. On the other hand, when the fishing goes well, the coastal man, by exchanging goods for fish, helps himself easily to everything he needs from the town. He lives rather well and best of all, yes even to excess.[1]

This system of valley, fjords and coastal areas which the author wrote about, is typical of the western parts of Norway. Regional differences in means of subsistence and living conditions were far stronger along the east-west axis, that is from the valley along the fjord to the coast, than along the north-south axis, for example, up or down the coast or from valley to valley.

The author cited did not invent this grouping of the population into valley people, fjord people and coastal people. These were concepts used by people in the regions about themselves and others till far into our century.[2]

Applying an ecological perspective on this regional differentiation, the valley, fjord and coast will constitute three different ecosystems. The differences between them are based upon the uneven distribution of energy sources. Most of these can be grouped within the activities of agriculture, forestry or fishery, according to the various different niches which people exploited in connection with these activities. Furthermore, in order to illuminate, for instance, the more uncertain living conditions of the coastal people, we must also bring in the concept of the periodicity of the energy sources, that is when, and how often, they can be exploited, and also how reliable they are from time to time. Schematically, the characteristic traits of the three ecosystems can be represented in this way.

ENERGY SOURCES CONNECTED TO:	IMPORTANCE IN ECOSYSTEM:	PERIODICITY:
Agriculture	Valley Fjord (Coastal)	Stable, yearly, controllable
Forestry	(Valley) Fjord	Stable, yearly, partly controllable
Fishery	(Fjord) Coast	Unstable, unpredictable, beyond control

This table gives a picture of how the peasant households in the valley and along the fjord utilized stable and fairly controllable sources of energy connected to agriculture and forestry. Even if they had to calculate with elements of risk, such as crop failures, cattle diseases, forest fires and other natural catastrophies, they lived in environments that made them less vulnerable to changes from year to year.

In the coastal region, on the other hand, which was woodless and where the agriculture was more marginal, we find an ecological adaption that was built, to a high degree, upon energy sources which were unstable, unpredictable and beyond control. The fisherman-peasant households along the coast had to calculate on wide fluctuations between plenty and want. This applied especially to the important seasonal fisheries in pursuit of spring herring (late in the winter), and of cod (later in the spring). The herring was most unpredictable. For years it would enter the coast to spawn in abundance, and then disappear for decades.

One of several ways for the coastal people to counteract such unpredictable fluctuations, was their remarkable ability to adapt to the changes, to find new niches to exploit. But with their one-sided dependence upon energy sources from the sea, they also had to find other ways.

As mentioned in the above quotation, they did not only live off the fish and whatever their agriculture would yield. They were also involved in trades of different kinds to get what they wanted or needed. To understand this, we have to turn from the ecological perspective to that of the economy. The coastal people were food-producers, involved in the production and distribution of perishable goods. So it is of great importance, for the understanding of their economy, to lay bare the various aspects of the conservation of their main products, fish and herring.[3] I shall, in what follows, go deeper into this by using examples from the economic adaptation of the coastal people.

In the perspective of economy, the energy sources become resources. It is important to stress the point that there are no resources as such, but only possibilities of resources provided by nature, in the context of a given society, at a certain moment in history.[4] It is thus not only in the natural environment that one has to search for those factors that decide which resources a given society exploits, and in which way they further use it. The relations of the society at large also play a part in forming patterns of subsistence.

The legal rights of ownership and usage of many resources influence strongly the social distribution of these resources.[5] From this point of view, we can uncover another important difference between agriculture and forestry, on the one hand, and fishery on the other. The resources exploited within agriculture and forestry were controlled by patterns of ownership and usage rights connected either to the single peasant, or crofter, household, or to the local community. Fishery resources, on the other hand, were mostly free for anybody to exploit, as long as they had the right knowledge and the necessary equipment. This led to another element of competition for the fisherman-peasant households along the coast. They did not only have to compete with an unpredictable nature, but also with strangers, visiting fishermen, for the resources of the sea. This element of competition manifested itself not only in the production (the hunting of the fish, so to speak) but also in the further distribution of the catch. Because they were entirely dependent upon the resources from the sea, the coastal people had to be exceedingly strategic in the planning of the distribution of the catch. This applies to their own consumption, as well as to the sale or exchange.

What role did the conservation process play in the economic strategies of the coastal people? To answer this question, I shall begin with a table formulated by Jack Goody:[6]

PROCESSES	PHASES	LOCATIONS
growing, hunting, gathering	production	farm, sea, wood
allocating/storing	distribution	storehouse/market
cooking/eating	consumption	kitchen/table

The middle column presents the abstract concepts of production, distribution and consumption. They are the three sub-structures of the economy. But when we are dealing with food in this perspective, they will also be phases in providing and transforming food from raw material to a meal. Some less abstract processes and their appropriate locations are connected to this pattern in the table.

It is obvious that the process of conservation must be the final stage in the production and preparation processes of perishable food. However, the conservation process goes beyond that. It enters the distribution phase, as the food is placed in the storehouse for later allocation. This is distribution in time, as food must be kept in stock between the productive periods of the cycle of the year. Thus conservation is a prerequisite for the allocation of the food. The consequent process of allocation has, from the point of view of the household, two directions. One is inwards, as the food is distributed for direct consumption through the year. The other is outwards, as the food is sold or exchanged on the market, for others to consume.

This broad interpretation of the conservation process reveals the importance of considering stocks of food in the strategies of economic adaptation of pre-industrial food producers.

By examples drawn from the region of Sogn og Fjordane, I shall penetrate beneath the surface of these processes and phases, starting with the production.

The relationship between the conservation of food and the work prior to it is most obvious where one and the same production group organise and carry out all the processes. This was the case in most places in pre-industrial agriculture, before the era of commercial slaughter-houses and dairies. Members of the peasant household took care of the different tasks themselves, all the way down to the storing of the conserved food. Thus the further distribution of the food was fully in the hands of the household.

The division of work could be more complicated in other kinds of food-production. Moreover, the further distribution of food could be beyond the control of the food-producers. This was often the case with fish, where the composition of the production unit, the fishing crew, might change according to what kind of fishery and what kind of fishing gear was relevant. The control of the further distribution depended on the fishermen having the capacity to organise and carry out the necessary conservation. The alternative was to sell the fish raw. The distance between fishing-ground and home was also important.

In the fisherman-peasant households on the coast of Sogn og Fjordane, the male members were engaged in fishing for the greater part of the year. This was mostly fishing in the inshore waters near home, an activity with which other necessary tasks, especially in agriculture could be accommodated. This fishing with handline, longline, or nets was carried out by one man, or only a few men, in the boat, mostly from the same household. Thus they provided the household with fresh fish and at the same time made it possible for the households to build up a store of conserved fish, salted, dried, smoked or a combination of these methods.

The big seasonal fisheries were a different matter. Few of those taking part lived near the spawning grounds of cod and herring. The fishermen, organised in bigger crews, left home and partook in intensive fishing through the season. In the pre-industrial fisheries, the level of technology was such that the fishermen were forced to go ashore each evening with the catch. Further north in Norway, this circumstance was exploited by the so-called *væreiere*, landowners near the fishing grounds, who were involved in trade with fish and other goods. The fishermen were allowed to land the catch on their ground on two conditions: the sole right to buy their fish raw and the sole right, as part of their payment, to sell them whatever provisions they needed. The main foundation of the economic activity of the *væreiere* was thus the trading of the conserved fish.

In Sogn og Fjordane, however, the shores near the seasonal fishing-grounds were available free for anybody wishing to land his catch. Here, the fishermen had the options of selling their catch raw or providing for its conservation themselves. Most of the fishermen in the coastal areas of the county chose the latter solution. They made provision for conserving their own catch and sold off raw any surplus which their own production apparatus could not deal with. The established buyers in the district had to rely upon the fishermen from other parts of the west coast partaking in these fisheries, since they most often chose to sell their catch raw. This gave them a greater mobility, which could be an advantage, especially when dealing with the unpredictable herring.

To illustrate these relationships, the spring herring fishery in the 19th century will be used. In 1808, the herring returned after having been away from the shores of the west coast for about 30 years. From then on, they came back each year until the 1870s. In this period, they gathered off different parts of the coast every year, but one of the most regular places was the archipelago of Kinn, in the outermost parts of Sunnfjord. In the first period of this herring fishery, the fishermen stored the herring raw or slightly salted in the open air.[7] After the end of the season, they brought this herring back home. There it was properly salted in wooden barrels for further use. Although it was winter time, and the weather was fairly cold, the herring was more or less fermented. This led to complaints from authorities on behalf of the consumers. The lack of proper conservation facilities near the fishing ground was only one of the reasons for this procedure. The most important reason was the lack of competition on the market. (There was also a very old local tradition which gave preference to fermented herring with a sour smell.)

This procedure changed in the next period. As the herring fishery continued steady for many years, more and more people came to rely upon the fact that it was more than a short-lived phenomenon. The herring also came in larger and larger quantities. From the 1830s, buyers established themselves on islands and skerries near the fishing-ground. Here, they built herring salteries, large two-storey buildings with living quarters for visiting fishermen and herring-salters on the first floor, and saltery and storehouse on the ground floor.

The buyers did not partake in the fishery. They were local merchants, citizens from Bergen and Stavanger and well-to-do peasants from the districts, mostly fjord peasants owning forests. They had the resources both to construct herring salteries and to produce the wooden barrels which were needed in large quantities. The buyers employed seasonal labour to do the salting of the herring, mostly young women from the towns. It was difficult to employ coastal women, because they were employed in handling their own herring. The buyers also had to rely upon visiting fishermen, who might disappear from the district in the years when fishing was better in other places. Then the buyers protested loudly, because the local fishermen were not willing to sell even one sour herring.[8]

The fisherman-peasants started establishing their own smaller herring-salteries as an answer to the competition from the buyers. As a result larger amounts of herrings were salted when fresh, and the market increased the demand for salted herring of high quality.

In 1862, there were 409 herring salteries in the county of Sogn og Fjordane. More than 80 percent of these were situated in the archipelago of Kinn. The number there was 339, of which the citizens owned 107, the local merchants 16 and coastal people and fjord people 214. A rough estimate shows that more than half of these were owned by coastal people.[9]

In 1850, however, an estimation[10] was given based upon statistics of herring traded in the cities which amounted to 60,000 barrels. This estimation indicated that between 10,000 and 15,000 barrels of herring produced in this area were not sold in cities. Quite a large amount of these 60,000 barrels were shipped to Bergen by the coastal people themselves, thus playing a very important part in their relationship with the market economy and the big firms in town. The 10-15,000 barrels being withheld by the coastal people were probably meant to be used for exchange on the local market and to be part of their subsistence economy.

This example points to the importance of viewing the conservation of food in connection with the prior processes of producing it. This is a point often forgotten in monographs dealing with fishing as an economic activity. The process of fishing is treated with great care; the process of conserving the fish is hardly mentioned. As seen from the point of view of the peasant-fisherman household, the conservation process is not only connected with the production process as a means to prolong the durability of the food; it is also connected with strategic planning and securing the viability of the household through the year. The storing of salted or otherwise conserved fish and herring gave economic security from one production period to the next. Here we enter upon the phase of distribution and the connected processes. I shall start with the inward movement, the distribution within the household, and proceed with the outward movement involving market relations.

The dependency upon conserved fish and herring in the coastal households expresses itself most clearly in their everyday meals. Three out of four meals a day involved fish or herring of some kind, starting with breakfast. The vegetables used were mostly potatoes, barley and oats. Meat was eaten on Sundays, if available.

The woman in charge of the allocation of food had an important position in the household. That is why the other members often called her 'food-mother'.[11] One of her tasks was to measure out the food in store according to both the daily needs and the possibility of obtaining more food. She had to have control over the demand for food in relation to the productive and unproductive periods of the yearly cycle.

This short term cycle must be seen in the context of the long term development cycle of the household. In this, the unproductive persons of the household were related to the productive ones. The producers were obliged both to those who in time would be producers, the children, and those who had been producers, the old ones. One of the most important obligations was to give them food, thus ensuring the physical reproduction of the household. In Norway we had an institutional form for taking care of the elders, which gave them rights either to partake in the meals or to receive specified kinds and amounts of food.

In addition to the family members, servants were usually employed in periods when the number of producers in the household was low. The most important emolument for the servants was their right to meals through the year. In these parts of Norway, the social difference between the peasants and their servants was smaller than in the richer agricultural regions. They shared the table and the meal. One of the ways the servants in these parts evaluated their masters, was whether they provided all meals or not. As one observer mentioned,[12] they were careful in this respect. If a servant had to skip a meal for one reason or another, he would demand two meals the next time food was served.

As has been mentioned earlier, the households of the coastal people were not self-sufficient in all the foodstuffs they needed. Their agriculture was poorly developed and they had to rely to a large extent upon the outside world for supplies, including grain, meat, potatoes and milk products. In addition, they needed wooden products such as barrels, salt, materials for fishing gear, etc. For items of trade, they had to rely upon the fish and herring in store.

These relations to the outer world were not accidental. They followed a fixed pattern from year to year with clear objectives as to what could be got and from where, and which goods were expected in return. Trading took place at certain times of the year, in relation to what kinds of goods were in store, thus following the production cycle of both parties. Thus the coastal people were involved both in reciprocal exchange based upon principles of the peasant economy, and in market exchange under the influence of commercial capitalism.

The reciprocal exchange relations of the coastal people can be subdivided in two. First, there was the exchange of goods with the fjord people and partly also with the valley people. Second, there was the yearly visit to the local fair or market-place. The means of communication was the boat, and the fjord was the highway. Most often the meeting-place for the fjord, valley and coastal people was half way up the fjord. The goods that were most important for the coastal people to get were: potatoes, sour milk, butter, cheese, flour, firewood, bark used for impregnating the fishing gear and different tools made of wood. They brought with them: fat herring, coalfish, redfish, cod, ling and cusk, all salted in barrels.

This exchange took place several times a year. The most important time was just after Michaelmas, at the beginning of October. At this time, the harvesting was over, and the coastal people had salted fish from the summer fishing in stock.

The other kind of exchange took place at the market-places, of which the largest was at Laerdal at the inland end of the Sognefjord. It lasted for eight days at the end of September. Here the coastal people met with others from a far larger area. Peasants came from widely separated parts of eastern Norway to trade their grain with fish and herring.

These exchanges, which took place both along the fjord and at the market-place, were direct and immediate exchanges, which can be described as demonstrating balanced reciprocity. Each party was in need of what the other could offer, and the exchange usually took place without any money being involved. The most common goods were evaluated against each other, and this value was kept steady for years, free of the influence of market fluctuations in the society at large. One barrel of salted coalfish was, for instance, considered to be equal to four sacks of potatoes. A barrel of cusk had a somewhat higher value.

The contacts with the citizen-owned business houses in Bergen linked the coastal people to the capitalist economy. This contact involved other kinds of goods, and took another form. It was to Bergen that they brought most of the surplus from the seasonal fisheries, salted spring herring and cod, either air-dried or split, salted and dried. They brought back home with them salt, grain, hemp and other plant fibres, stimulants such as coffee, brandy, snuff and tobacco. And, when times were good, they could help themselves to some of the luxuries the town could offer, finer clothes, stylish furniture and such.

These journeys to Bergen took place a few times each year, just after finishing the conserving of the fish and herring. They used specially built ships, the so-called *jekter*, with a carrying capacity of several hundred barrels. These ships were often the joint property of all the households in a community, who cooperated in equipping them, manning them and using them.

The relationship of the citizens of Bergen to the fishermen-peasants can be characterised as a patron-client relationship, thus quite asymmetric. Trade with the big business-houses always involved money, though seldom cash. It was a system of credit, where account-books were used. As the fishermen-peasants were usually in constant debt to one firm, it bound them to the firm maybe for generations.

Another negative trait of this relationship was the link with the capitalist economy and the consequent following of its laws of supply and demand. Prices fluctuated from year to year, and even within a year. The coastal people could not predict the value of their goods in the same way as in exchange relations with other peasants.

Even if the citizens were the strong party in this relationship, there was, as in all patron-client relationships, a positive side, in this instance security for the peasants. Long-term contact with one single business house gave them economic security in hard times, when they could buy on credit and postpone the payment in goods till they were better off.

In this paper, I have used a broad interpretation of the concept of food-conservation. In the perspective of the economy, I have related it to the three phases or substructures of production, distribution and consumption, by the example of the coastal people of Sogn og Fjordane, dependent as they were on fishing. I shall conclude with a question. Can we, by focussing on food-conservation, gain new insights into the economic adaptation of food-producers? For my part, the answer is yes. The concept encompasses both the production, storing and allocating of food. Thus it grasps at one time the three different economic spheres in which the peasants were involved, i.e. (a) the subsistence economy, (b) the peasant exchange economy or petty commodity production, and (c) the capitalist economy. By doing this, it gives me a fuller understanding of the peasants as active people making strategies for their future.

REFERENCES

1. From *Søndfjords beskrivelse* by Hans Arentz, written about 1785, published after his death. Arentz, who was born in Sunnfjord and later became district stipendiary magistrate there, was a keen observer of the peasant life in the district. The quotation is from Amund Helland,*Topografisk-statistisk beskrivelse over Nordre Bergenhus Amt*, vol I, p 682.
2. As mentioned often in answers to questionnaires sent out by Norwegian Ethnological Research, Oslo. In this paper, I have systematically used questionnaire no 27, Fish as food.
3. The distinction between fish and herring may seem artificial from a biological point of view. But in popular practice, it is different, fish is fish and herring is herring. It stresses the important role of herring both in food and economic activity.
4. Maurice Godelier, *The object and method of economic anthropology.*
5. I have presented a more profound discussion of this aspect in connection to an empirical investigation of a typical fjord district. See Bjorn Fjellheim, *Sand a fore - singel a selge.*
6. In Jack Goody, *Cooking, cuisine and class*, p 37. I have modified his table in two ways. First, I have combined the two processes of cooking and eating, with their phases and locations, into one. This is because I conceive that the phase of preparation connected with the process of cooking in Goody's table belongs to a different analytical methodology than the other phases. Second, I have used more general terms for the locations.
7. Amund Helland, p 571.
8. Karl Egil Johansen, *Fiskarsoga for Sogn og Fjordane* 1860-1980, Universitetsforlaget 1982, p 52.
9. Arne Faeroyvik: *Vårsildfisket i Kinn* 1830-1870.
10. *Ibid.*
11. The Norwegian word *matmor* in direct translation.
12. Hans Arentz, *Søndfjords beskrivelse.*

LITERATURE

Fjellheim, Bjørn, *Sand à føre - singel a selge. En studie av okonomisk tilpasning i vestlandsbygden Stamnes 1870-1940*, unpub. thesis, U. of Oslo, 1982.

Faeroyvik, A.,*Varsildfisket i Kinn 1820-1870*, unpub. thesis, U. of Bergen, 1983.

Godelier, Maurice, 'The object and method of economic anthropology', in Seddon, David (ed), *Relations of production. Marxist approaches to economic anthropology*, London, 1978.

Goody, Jack, *Cooking, cuisine and class. A study in comparative sociology*, Cambridge, 1982.

Helland, Amund, *Topografisk-statistisk beskrivelse over Nordre Bergenhus Amt*, vol I, Kristiania, 1901.

Johansen, Karl Egil, *Fiskarsoga for Sogn og Fjordane 1860-1980*, Universitetsforlaget, 1982.

OUTLINE OF THE DEVELOPMENT OF FOOD PRESERVATION IN THE PAST HUNDRED YEARS

Jozien Jobse-van Putten

The decline of home food preservation over the past hundred years in the Netherlands, and certainly in the Western countries in general, can only partly be accounted for by the decline of the system of self-supportiveness. Recently, a renewed interest in home food preservation can be observed, independent of this system. Besides, a number of changes and innovations in preservation have taken place, even after the system of self-supportiveness had lost its previous importance. As is the case with the regional spread of the applied preservation methods, the changes can be accounted for partly by food-external and partly by food-internal factors. Although home food preservation is no longer a necessity but has become a free choice, its most important functions can still be recognised today.

INTRODUCTION

Food preservation at home has been characterized by a great deal of dynamism in the last hundred years. There have been changes not only in the techniques of preservation but also in the significance of preservation itself. In the course of the present century home preserving in the Netherlands seemed almost to have ceased, but today a revival is perceptible. This is the case not only in the Netherlands but also elsewhere in Western Europe. In recent years the various media have regularly turned their attention to food preserving, and the reaction to this shows that this media interest reflects the interests of readers and viewers. While a modern technique like freezing has become generally practised, all sorts of traditional forms of preserving are once again being extolled and are increasingly used in households. However, the significance of preserving, as compared with the situation at the end of the last century, has radically changed.

In this paper I shall examine more closely the changed significance of home preserving in the past hundred years. This will show that, although the extent of preserving and the techniques used have indeed changed in the course of this century, the most important functions of home preserving have so far remained constant, though in a diminished form. I have based my investigation on preserving in the Netherlands, and my examples are taken from there, but the main trends in food preserving will also be valid for other Western countries.

SELF-SUFFICIENCY

In the past home preserving was closely linked with the system of self-sufficiency. Until the last quarter of the 19th century the majority of the Dutch population lived in the flat countryside where the money economy was generally of limited importance. People tried as far as possible to look after their own provision of food. The seasonal variations in the supply of the products of agriculture and fishing made the storing of food necessary, and because most food deteriorated after a fairly short time these provisions had somehow to be given a longer storage life. So in the past the main function of preserving was to regulate supply and build up provisions of the most important foodstuffs.

My investigation of home preservation in the Netherlands in the last hundred years has revealed two important aspects of self-sufficiency. First, it appears that at the beginning of the present century there were great differences between various regions in the extent of home preserving. Nobody will be surprised that less was preserved in the towns than in the country, although preserving in towns was not yet a phenomenon of the past. But it was surprising to find that there were also extensive rural areas where preserving was not much practised. It appears that this was related to the agricultural structure of the regions, which determined to a considerable degree the possibilities of self-sufficiency in the past. Thus in the traditional livestock-rearing areas food preserving was of fairly limited importance; it appears that there it was largely left to the peasant farmers. In the other rural areas where arable and mixed farming predominated, almost all groups in society tried to preserve their own food as far as possible. This regional difference in the amount of preserving is no doubt connected with the fact that in the livestock-farming regions people are traditionally more trade-orientated and hence more involved in the money economy.

Secondly, it appears that the end of the system of self-sufficiency which occurred in the last hundred years proceeded more or less straight-forwardly for meat (although here there were also regional differences); while for fruit and vegetables there was first a revival of the system of self-sufficiency. As regards meat the reduction in home slaughtering (and the home preserving connected with it) can be regarded as a development which had begun before the turn of the century. In some regions home slaughtering had almost completely disappeared by the first quarter of this century. Elsewhere the tradition of home slaughtering and meat preserving seems to have been so strong that it continued to be practised by more than half the population until after the Second World War, and it was not until the 1950s and 1960s that it came to a fairly abrupt end.

In contrast to meat the preservation of fruit and vegetables generally increased in the first half of this century, even in regions where the system of self-sufficiency was a weak force. In the towns too around 1950 the number of women who preserved fruit and vegetables was fairly high. On the whole the movement from the countryside to the towns did not bring about a sudden break. People often brought the habit of preserving food with them, even though self-sufficiency had frequently become impossible and was no longer a necessity. Although they had joined the money economy and mostly had to buy the produce for preserving, to begin with many women still considered the building up of a store of food as part of their household duties—although this store was not as large as it had been. This identity function of food-preserving, which contributed to the self-realization of women (I shall return to this point), meant that preserving was not just a feature of the flat countryside but also of urban society and, once freed from the system of self-sufficiency, it displayed a certain amount of tenacity. Not until the 1950s and 1960s did the preserving of fruit and vegetables at home undergo a rather sudden and sharp decline. This was particularly the result of the modern farming methods and industrial food production which, combined with modern means of transport and retailing, could now guarantee the food requirements of Western countries for the whole year. Moreover, prosperity had risen to such a level that the system of self-sufficiency gradually became superfluous. I shall come back to the question of the consequences of this for home preserving.

The increase mentioned earlier in the preserving of fruit and vegetables in the first half of the present century can be largely attributed to the increase in the consumption of fruit and vegetables. Around 1900 these were eaten only in very limited quantities by a significant proportion of the working population in the towns and in large areas of the flat countryside of the Netherlands, and therefore the preserving of these products was not very extensive. In many regions home preserving only extended to two sorts of vegetables (usually green beans and sauerkraut) and one or two varieties of fruit.

In the course of the present century a number of educational and advisory bodies have disseminated health information in the towns and countryside to increase the consumption of fruit and vegetables. People have not only been taught to eat adequate amounts of fruit and vegetables but also recommended to have a greater variety. In the countryside people were given training and advice on the care of vegetable gardens and the growing of many varieties of fruit and vegetables, and thus self-sufficiency in a particular context, could increase in a period when the overall system of self-sufficiency was vanishing. Since, moreover, buying power was also increasing in this period, the number of kinds of fruit and vegetables which were preserved at home around the middle of the present century had significantly increased. This had been made possible by the application of the new food-preserving technology of bottling. This brings me to the second part of my paper: the changes in food-preservation techniques.

EXPLANATORY FACTORS; FOOD-INTERNAL AND FOOD-EXTERNAL

Although the system of self-sufficiency has been in decline in the present century, food-preserving was still in full swing. New techniques such as bottling and freezing spread rapidly. Traditional techniques lost their importance, though for various reasons they sometimes continued to be used in households alongside the new ones. The question of why people alternated the methods brings us to a large number of explanatory factors which can be categorized according to region, type of household, pattern of meals, sort of product, etc. These explanatory factors can be divided into two groups: food-internal factors and food-external factors.

Regarding food-internal factors it appears that changes in the pattern of meals had consequences for the techniques of preservation. Thus the transition from stews to a meal of potatoes, vegetables and meat with gravy, all cooked individually and served individually (though side by side), had consequences for the use of salting as a method of preservation. Traditionally salted meat, for instance, was often cooked in stews so that the disagreeable salty taste was reduced. In the new type of hot meal the inclusion of salted meat was problematic if it was eaten on its own, it was too salty and a tasty gravy could not be made from the liquid. The introduction of the new type of hot meal therefore led either to the end of salting meat or to another use for salted meat.

Conversely, changes in methods also had consequences for the pattern of meals. The advent of the bottling not only meant that this new method replaced old techniques of preserving, it also created new possibilities. To begin with bottling was used particularly for those varieties of vegetables which had previously been rarely or never eaten, such as peas, little carrots and cauliflower. Because bottling, unlike traditional preserving techniques, was not linked with specific flavours, bottled products were distinguished from other preserved foods as 'fresh' and for a few decades were often treated as food for Sundays or celebrations.

Besides the food-internal factors there were various social developments which also influenced the preserving techniques used. I shall give a few brief examples. New scientific knowledge about nutrition, which gradually spread to the population through education and publicity, made salting and smoking seem methods with negative effects on health. Increasing prosperity brought relatively expensive bottling within the financial range of the less well-to-do population. The replacing of the open fire with the stove, largely because of a change in techniques of chimney construction and the transition to coal as fuel, made the smoking of meat at the domestic level no longer possible.

One of the most important of the food-external reasons which led in the Netherlands to a change in methods was, however, the technical aspect of preserving. The preserving of foodstuffs in households was until recently in actual fact an unsolved problem: none of the methods used was ideal. Every technique had its own drawbacks. Salted products, for example, became after a time too salty to be eaten. If food was smoked or dried it often became too hard after a short time, or if the humidity of the air was too high—as is often the case in the damp Dutch climate—it soon became rancid. Changing conditions and lack of knowledge about the preserving process meant that people were never sure about the quality of the end product or the length of time within which it could be eaten and preserve its flavour. Disquiet over the changeable quality of the early preserving can be regarded as the driving force behind the continual changes and experiments in techniques.

Although bottling in general led to better qualitative results, there were also disadvantages attached to the method. An important feature of traditional techniques was that they could usually be employed without using expensive sources of energy. Bottling on the other hand not only took a great deal of time and work, its equipment and fuel was also fairly expensive—an argument that had to be considered in the early period of bottling. Furthermore there was the risk that a bottle would go bad if it had not been treated hygienically enough or the instructions had not been followed exactly. Since bottling had spread mainly by word of mouth, this last danger was not an imaginary one. These drawbacks explain among other things why bottling was later replaced to a large extent by the new method of freezing.

The regional differences in the preserving methods used (see map) can also be explained partly by factors internal to food and partly by external factors.

The earlier cited example of salted meat (see Note) can also serve as an example of the food-internal explanation of regional variations in the methods. The lengthy salting of meat (a method not followed by a drying or smoking process) was significantly more prominent in some regions in the Netherlands than in others. This salted meat was usually cooked with stews, soups or pulses. Because in the regions where a part of the meat was merely salted, stews of this kind are eaten frequently (much more frequently than elsewhere), it can be assumed that the meal pattern and the preserving methods used must once have been interdependent.

One of the most important food-external factors in the regional differences in the Netherlands between smoking and drying meat is the structural development of the chimney. The smoking of meat, by hanging the meat in the chimney or in a smoking chamber was familiar throughout the Netherlands at the beginning of this century, but in some regions more than in others. The drying of meat, where after salting the meat was hung up from the ceiling, was only found in the east and, to a lesser extent, in the south of the country. In these regions the chimney was a relatively recent phenomenon. In the east where the Lower Saxon house type had long predominated, houses without chimneys persisted until the end of the last century. Because the smoke of the open fire moved freely round these houses smoke was always mixed in with the drying of the meat, which was thus indirectly smoked. When finally the chimney became the norm, the wide flue necessary for good smoking declined because of the increasing use of the stove and different fuels. In a small flue it was no longer possible to smoke produce in the chimney. The result was that people continued to hang meat from the ceiling as they had done in the past, and because the room was now smoke-free the method—wittingly or unwittingly—was changed: the earlier indirect smoking was transformed, without changing its manner or position, into the process of simply drying the meat. Consequently the regional differences between smoking and drying in the Netherlands are of relatively recent date and are to a large extent bound up with the developments in heating technology.

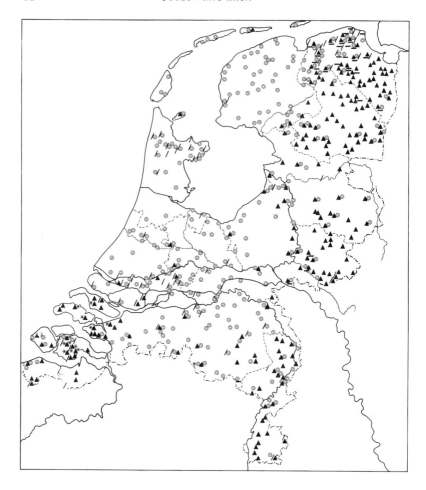

/   salted

▲   dried

◎   smoked

—   in buttermilk or vinegar

＊   under a top layer of lard

Regional preferences of the conservation methods of pig fat in the Netherlands.

THE FUNCTION OF DOMESTIC PRESERVING

Lastly I shall examine in more depth the altered significance of home preserving in the past hundred years. When the money economy was still of limited influence in the countryside food preserving was closely related to the system of self-sufficiency. It seemed reasonable to assume that preserving would vanish with the passing of this system. Now that a new method like deep freezing has become common both in towns and in the country independently of the system of self-sufficiency, there are many indications of a new interest in some of the old preserving techniques, I think it is necessary to look more closely at the various functions of preservation in earlier times and today.

Three functions can be differentiated in home preserving. The primary function of regulating supply and building up provisions mentioned earlier, is related to the overall feeding of a family household. Various motives can play a part in this. In earlier times people were generally forced by necessity to lay in a store of food to bridge the period between the seasonal production of food and its actual consumption. In places where money economy played a more important role such food could also be stored to make one less dependent on the external regulation of food supplies, or because of the financial advantages of such provision. Besides this primary function there were also two secondary functions: the culinary function and the identity function.

The culinary function is related to the meal in which the preserved produce was eaten. Almost all the traditional ways of preserving, where the original aim was to protect the food from deterioration, involved the addition of particular flavours. If such a flavour was positively rated it could become a secondary aim of the preserving process - sometimes even the primary aim. So it would no longer be a matter of merely preserving food, but of creating a particular flavour. This culinary function applies to the preservation of produce to achieve a particular quality and the addition of more variety to the diet by using preserved foods.

The identity function mentioned earlier relates to the person doing the preserving. In earlier times preserving was part of the standard duties of the housewife. Successful results in preserving brought her a certain prestige. Thus in the middle of this century women in many villages paid each other visits to have a look at the stored food.

These three functions of preserving did not usually occur separately; usually several of them were at work simultaneously. The significance of each individual function has changed a little in the course of time, as has their relationship to each other. When after the 1950s, for example, self-sufficiency seemed to be becoming unnecessary, the emphasis shifted. Thus the earlier distinction between the primary and secondary functions of preserving can no longer be made today. The three different functions are of about equal importance, although of course they still occur in combination with each other just as they did in the past. Now we shall examine how the significance of the functions has adapted to the present day.

With the passing of the system of self-sufficiency in the past hundred years what used to be the primary function of preserving - regulation of supply and building up of provisions - has lost its character of inexorable necessity; today it is based on free choice. If people no longer have to preserve food, but do so from free choice, they can weigh up the advantages and disadvantages. For instance they can ask themselves whether the financial advantages and/or the attainment of particular flavours are worth all the effort and risk involved in the storing of food. The identity function could also influence the decision whether or not to carry on preserving. People who did not enjoy preserving and thought that it involved too much hard work could give up self-sufficiency altogether. However, there were also women in the Dutch countryside who thought that they would be idle and not fulfilling their household duties if they gave up the vegetable garden, home slaughtering and preserving. Around 1980 there were still some elderly interviewees who proudly told me about the contents of their deep-freezes full of home-grown and home-preserved produce. Besides the function of building up provisions (today this is freely chosen), the identity function also seemed to be of great significance to them. The functions of preserving seemed sometimes to have survived longer than the traditional techniques, though the old preserving techniques have not completely vanished; they are occasionally still used today if, for example, people do not want to do without a particular flavour or a specific dish for which a particular preserved product is needed (culinary function).

Although home preserving in the country has lost much of its former significance and its extent has considerably diminished, preserving food there today is not just practised by a few elderly women. In many young households too there is a deep-freeze which is sometimes filled partly with industrially frozen food, as well as with home-grown produce. In many regions the tradition of having a vegetable garden rather than a flower garden is still quite strong. The owners of vegetable gardens almost necessarily have to preserve their produce because there are always surpluses at harvest time. Although we are dealing here with a certain kind of self-sufficiency, these instances cannot really be regarded as continuing the old system of self-sufficiency. The system of self-sufficiency belongs to a particular type of society in which shortage of money, transport, a good system of food supply and such like made it necessary for people to look after their own requirements. When people today still preserve food in the countryside they rarely if ever attempt to lay in complete provisions for a whole year. They preserve only what there is too much of. The remainder is sold—either fresh or preserved. In this way the negative effects which used frequently to be linked with the keeping of long-term provisions are avoided. But the former function of regulating supply and building up provisions is still recognizable.

In the towns too the function of regulating supply and building up provisions has been rehabilitated either out of free choice, outside the agricultural context and adapted to today's social conditions. The rise of the supermarkets, the transition from weekly wage to monthly salary, the possession of a car and the increase in the number of working married women have caused people to purchase a large store of food for a longish period, and this of course involves preserving. The argument that one will thereby be able to have particular products available throughout the year is reminiscent of the traditional primary purpose of preserving. In these instances preserving as such no longer needs to be carried out on a domestic basis; the housewife only has to cope with the organization and control involved in building up provisions.

The use of food preserving as a form of leisure activity can be regarded as a new form of the identity function; this is not tied to the town or the countryside. In shops the complete apparatus for bottling is available and a large number of cookery books encourage the preserving of vegetables, fruit and sometimes meat and fish, using both traditional and modern methods. The necessary utensils are in the shops: various sorts of pots, small smoking vessels, sawdust for smoking and so on. In the field of leisure activity all kinds of earlier forms of preserving, which had been overshadowed by the rise of new techniques and the decline of the necessity to preserve food, are reappearing. Elements such as creativity, nostalgia and sometimes a striving after a particular quality play a part in this.

This brings us to a recent variant of the culinary function: when food is preserved for the sake of creating particular flavours. Some groups in the population also do this because they require purity in food, the result of an increasing dissatisfaction with factory production in which all kinds of chemical additives are used. Also those who like frequent variety in their food are in favour of preserving produce at home. The daily supply of food is thereby increased and the possibilities of introducing variations in taste are almost unlimited.

CONCLUSION
It can be concluded that the home preserving of food which was formerly linked necessarily to the system of self-sufficiency can nowadays also function independently of this system, but to a far more limited extent than in the past. It appears that food preserving is not an extinct cultural phenomenon. It has adapted to changed conditions and even displays a quite remarkable vitality.

NOTE
The word 'meat' in following paragraphs is used in a very general sense. It includes cuts of meat, ham and bacon, as well as sausage and so on. The map showing the distribution of preserving bacon should be seen as an example of one of the various sorts of meat.

BIBLIOGRAPHY
Jobse-van Putten, Jozien 'Veranderingen in de huishoudelijke vleesconservering in de afgelopen eeuw', in *Volkskundig Bulletin,* 10 (1984), pp 1-49.
Willinge Prins-Visser, C.W. 'Conserveren van levensmiddelen op het platteland', in *Mededelingen van de Landbouwhogeschool te Wageningen,*56 (1956), no 2.

PRESERVING FOOD AT HOME OR BUYING IT TO HOARD? The Organization of Stockpiling as a Means of Cultural Expression

Christoph Köck

This paper sets out to demonstrate the significance of the way stockpiling is organised in various cultural milieus. Changes in its organisation are not just the consequence of new technology and economic conditions.

All those who came to this conference in Sogndal by sea and land must have noticed on the ferry that brought them from the European Continent to Scandinavia a characteristic institution found only in the no-man's land of ships or airports. I mean the duty-free shops, those much appreciated tax refuges between nations.

You will all be familiar with the scene from your own observation or participation. Shortly after the ferry casts off a large crowd of passengers gathers at the still closed doors of the duty-free shops and awaits the steward's starting pistol. Once the ship has left national waters the shop door finally opens to reveal consumer goods of all kinds: cartons of 200 cigarettes ten marks cheaper than at the kiosk at home, Pernod two marks below the discount price, jelly beans and coffee at special rates.

We Germans are not the only ones to grasp this opportunity to lay in provisions; the Scandinavians, who are particularly badly hit by alcohol restrictions and high food prices, do so too. There are import resrictions to prevent foodstuffs being hoarded in excessive quantities. But both the Germans and the otherwise law-abiding Swedes feel their conscience wavering between legality and luxury, and many regard outwitting the viligant customsmen as a sort of forbidden game—one in which discovery entails severe penalties.

The *Sydsvenska Dagbladet* of 1 December 1984 describes the day before the duty on alcohol was raised (yet again): 'This rush can no longer be called mere hoarding. These were super-hamsters. ... A young man bought precisely 10,000 kronen worth of alcohol. All of it—from Spanish country wine to the most exclusive brands of whisky—by the crate.'[1]

The list of examples of the stockpiling of provisions for a great variety of motives is endless. Germany in the 1970s had its very popular 'butter trips' on which one could have an enjoyable day out and buy a lot cheaply. In the months following Chernobyl uncontaminated dried milk powder was at the top of the list of goods demanded by consumers and was consequently hoarded. Some of you will know from personal experience the scenes in Germany in the period after the war when the city population swarmed into the surrounding farmland to alleviate the shortages in the conurbations by stealing potatoes or begging—although these were forbidden.

Situations that occur quite naturally today are reminiscent of features of hoarding in times of shortage: the bulk packs and giant quantities that we buy at the supermarkets outside our towns fill our cellars and deep-freezers to the brim. We enjoy showing friends, relations and neighbours the wine bottles lining the racks in the cellar. Their contents have for long merely made a pretence of fulfilling their function as emergency food supplies. But we must never be in the embarrassing position of having nothing in the house.

Can this stockpiling of provisions be explained simply as the result of the bitter post-war experiences and anxiety about future crises?

The relevance of my paper to the theme of the conference lies in the fact that the organization of the stockpiling of food reflects not only technological knowledge and the nature of the processing methods used; it also tells us something about how people organize their provisions and the various cultural scales of values which underlie the acquisition and storing of food.

It is therefore interesting, for instance, to combine Wiegelmann's model of the 'Meal as Central Complex in Ethnological Food Research'[2] with the analysis of culturally specific concepts of values. Using this model we can ask what role social, economic and technological changes play in the organization of the stockpiling of food. Besides this we can trace whether these changes are expressions of particular ideational processes. How has the combination of new concepts of values and changed external conditions affected the organization of food provision?

My discussion of this can be divided into two parts. I shall begin by examining the cultural significance of preserving and storing food in various social groups during the period of the 'revolution in household technology'. By comparing and contrasting farming, home-industrial and bourgeois sections of the population I shall attempt to identify various concepts of values underlying the way each of these groups organize their food supply. The second part of the paper is centred on an attempt to discuss the change in values implied in the diminution of home-made provisions in bourgeois society. I should also like to sketch how and why the field of work in 'self-sufficiency and food-preserving at home' has been increasingly broken down, and how the storing of provisions is determined by 'outside food-preserving and shopping'.

THE SIGNIFICANCE OF STORING PROVISIONS BEFORE THE HOUSEKEEPING REVOLUTION

People ignorant of modern technological methods of food preservation had to store provisions at fixed times of the year. The economy of the region determined the nature of the products that could be preserved and stored, and the amount stored was directly dependent on the size of the harvest and the products of animal husbandry. These yields were affected by short-term events in the natural environment and climate. Poor harvests were often the result of natural disasters such as periods of drought or rain, and the ultimate consequence of these was famine. Methods of preserving and storing food were necessary in pre-industrial German society to stay alive during the unproductive months of winter and spring.

Among the population of peasant-farmers the self-sufficiency associated with storing provisions was rooted in a cyclical consciousness of time. The farmer's way of thinking, planning and dealing was determined by the constant round of the seasons. The range of provisions did not have much of a representative function: around 1900 farming families in south-west Germany still bought only very few products - such as sugar, noodles, salt and coffee - at the market. Most of their stored provisions consisted of the products of their own farming. The result of this long-term stockpiling was a monotonous diet.[3] As a rule home-preserved foodstuffs were eaten on everyday occasions and were less prestigious than the freshly killed or harvested products which appeared on the table at celebratory meals. Meals made up of stored food had connotations of monotony.

Wiegelmann emphasizes that in the 19th century the farmer's wives modified this monotony by using a strategy of variation.[4] They organized the provision of food by forward planning and this was a means of gaining prestige. The important thing was to vary meals for different occasions (for normal everyday eating and celebratory feasts) and for different seasons. It was not the richness and multiplicity of the stores of food accumulated at particular times that was valued among the farmer's wives, but the way in which the food available was treated. Preserved provisions were integrated by the farming population into a system intended to regulate essentials and so obviate shortages. However, this system was easily upset: shortages resulting from social injustice, political events, sickness and epidemics were rather the rule than the exception. Also long-term food storage was not always as farsighted as is assumed. In the early modern period in particular 'gluttony and drunkenness' were widespread at festivities, when 'enormous quantities of food and drink were consumed and everything laid by was used up'.[5]

Since the 18th century the farming population was confronted with a way of life that differed strongly from its own as home industry spread throughout many rural parts of Germany and the rest of Europe, and cultural changes were apparent in these regions. The nucleus of the home industrial workers were people who had previously belonged to the lower stratum of peasant-farmers and who, because of the rising population, had been deprived of secure employment. They emancipated themselves from the habits of a farming life based on self-sufficiency. A money economy and market-oriented consumption were dominant features in the world of the rural home workers. Time became money and the possession of money was the basis of providing for oneself. All the members of a family would work to earn money, and because the former housewife was now at the centre of the production process and could spend only the minimum of time on housekeeping, the storing of provisions was scarcely feasible. The culture of the home workers involved demonstrating their independence from their former financial constraints. Instead of preserving food they made frequent trips to market and bought 'the delicacies of the townsfolk, ...coffee with the richest cream or veal at the times ...when it is at its rarest and dearest' ['die Leckereyen der Städter, ...Caffee mit dem fettesten

Rahm oder Kalbfleisch in solchen Zeiten, ...wo es am selthensten und theuersten ist'], wrote a contemporary in 1790.[6] The habits of the home workers were less affected by cyclical conceptions of time and broke with the pattern of peasant-farmers' lives. The rejection of the orderly regularity of peasant culture was symoblized by their luxury consumption and at the same time brought about the construction of a cultural pattern of their own. The farming population defended their traditional way of life. They criticized the rural industrial workers for 'not having the housekeeping frame of mind'. Their 'hand-to-mouth way of living' annoyed the traditional country dwellers.[7]

The shortcomings of the home workers' system of food-provision are evident: from now on what had been the lower stratum of farmers was dependent on industrialists, above all the textile manufacturers. If the laws of the capitalist market meant there was no work, the home industrial workers found themselves on the streets with no bread and no security. They could not fall back on the stored food or savings which the farmers laid by for crises. The culmination of the social crisis among the home textile workers was the Silesian weavers' revolt of 1848 against the dramatic decline in their earnings and living standards.

The influence of the markets also affected the style of consumption in the ever-growing urban regions. Upper bourgeois housekeeping in particular had by the 18th century changed from the medieval and early modern self-sufficiency in food towards dependency on food from outside.

Although this group always had capital it did not practise the sort of short-term provision that the home workers were compelled to adopt. Egner provides documentary evidence that 'many consumer goods were obtained in bulk from shopkeepers or directly from the producers' so as to remain independent from the annual cycles and from political and economic troubles.[8]

The housewife played the leading part in the storing of provisions. In the patriarchal system of the bourgeoisie the husband looked after public business while his wife, assisted by her many domestic servants, controlled the private, less demonstrative sphere. The Jewish merchant's daughter Fanny Lewald described the organization of storing of provisions in her household at Königsberg in 1825: [Housekeeping holds to the basic principle] 'that it is an advantage to buy in bulk where the cheapness of space makes storage possible; but one entertains the impractical desire to manufacture everything possible oneself at home'. Nearly everything was preserved and stored: basic foodstuffs, fruit, vegetables, spices and meat products.[9]

The organization of stockpiling in the upper bourgeois household reflected not only economic and patriarchal ideas. Preserving and storing provisions became a symbol of bourgeois notions of order and cleanliness. With a new delight in prettiness, people now took pleasure in the aesthetic qualities of storerooms, which had earlier been scenes of disorder. The contemplation of larders became a passion. The same merchant's daughter wrote:

stocking up with provisions far in advance was a pleasure, if one had the means to do so. The full larders and cellars with their stoneware jars, barrels, chests and drawers were a pretty sight. The dried fruit on strings, the marjoram and onions combined with the spices to give the larder a wonderful fragrance, the plump vegetables in the cellars smelt splendid. One had a feeling of contentment now that everything was gathered together. Now winter could come and do its worse![10]

Bourgeois households set great value by their image, and this was what differentiated them from the farming population. It was not the art of forward planning that brought prestige, but the quantity, variety and aesthetics of the preserved and stored food. The storing of provisions for winter and times of crisis is clearly a pretext for the bourgeois system of ideas regarding order and cleanliness. The establishment of long-term stockpiling also symbolizes their detachment from the urban lower classes who had to buy their food—if they still provided for themselves—all year round at the market and from shops. They never owned enough capital to lay in large supplies of food. Independence from the institutions of market and shop, except for a few bulk purchases, was a characteristic of the bourgeois housewife and contributed to her importance, as a final quotation from Fanny Lewald demonstrates: 'Anyone who expected [the housewives] to obtain their bread from the baker, their dried fruit from the grocer, their requirement of salted meat from a butcher, was regarded by them as a heretic, a sinner, who wished to restrict their housewifely duties ...'.[11]

THE CULTURAL SIGNIFICANCE OF NEW FORMS OF ORGANIZATION AFTER THE TURN OF THE CENTURY

This outline of some examples of the way the stockpiling of food was organized in three population groups before the revolution in household technology is intended to demonstrate the interconnection between the complex of 'buying-preserving-storing' food on one hand and the variety of cultural values. We can now ask how the values of these groups continued to affect these groups in the modern industrial age, and what new patterns of thought and practices are characteristic of the organization of stockpiling in this period.

Wiegelmann and Teuteberg emphasize that self-sufficiency in German towns was finally destroyed by provisions from outside only at the time of the foundation of the Bismarck Reich or the turn of the century.[12] The 'revolution in food-preservation'[13] is thus bound up with the process of the industrialization of food production: 'scarcely anything edible reached a town household that had not gone through a technical production process'.[14]

The new methods of preserving food made it possible for shops and markets to become public larders offering a wide range of foodstuffs all the year round. In a tin can - and later in a refrigerator and deep-freeze - food remains edible for a long time, and it is understandable that for this and other reasons many people changed to short-term food storage out of practical considerations.

In Germany since the 1960s the new household technology has generally diminished the importance of the rhythm of the seasons even in rural areas.[15]

The reduction in 'home-preserving' coincided with a new field of activity: shopping now becomes very important and is primarily the responsibility of (house-)wives. Those town-dwelling women who had long been accustomed to themselves treating the products of autumn so that they could be stored, gradually lose a large part of their prestige now that food-preserving is institutionalized. Images of the Wilhelmian period demonstrate how much bourgeois men liked to see their spouses in the role of Ibsen's Nora. The revolution in household technology meant that housewives were deprived of part of their responsibility. A contemporary wrote:

> Our dear housewives today do not know, or at least do not consider, how easy it is for them to run a household today compared with their grandmothers. It is no longer necessary to stock the house with foodstuffs in the autumn, one can buy them all through the winter from various shops as cheaply as if one had preserved and stored them oneself. ...Was it therefore to be wondered that, with all her household duties, grandmother found little time for reading books, playing the piano or embroidery?[16]

The task of the bourgeois housewife was now more a matter of busying herself with the pretty, homely decoration of the house. While her husband had to cope with the 'harsh' reality of the business life, the housewife had to withdraw to the private sphere and create a sympathetic atmosphere there.[17]

A comparison with the 1970s shows how enduring this division of labour has been and how the practice of shopping legitimizes it.

In a book of housekeeping advice published in 1971 we learn that 'shopping is one of the housewife's most enjoyable housekeeping activities. It gives her the chance to get a bit of fresh air, meet an acquaintance and have a chat with her.'[18]

Shopping thus replaces food-preserving as a part of the individual food supply chain, but it would be too simple to interpret this merely as the consequence of technological and economic changes. Norbert Elias's theory of civilization provides another approach to an explanation.

For people in 'civilized' countries the preservation of food outside the home is related to their more narrowly conceived schema of cleanliness. Food-preserving outside one's own four walls reflects the increasing tendency to place unpleasant or nauseating activities and objects in institutions beyond one's boundaries. Since the turn of the century the filth, blood and evil smells associated with the slaughter of animals at home have largely been consigned to communal or private slaughter-houses.Handling the bodies of killed animals, once a natural part of preserving food at home, was now left to professional specialists. One reason for this development is the increasing interdependence and interaction of people in a more strongly differentiated society.[19] Ideas of cleanliness determined the bourgeois way of life at the turn of the century even more strongly than at the beginning of the 19th century. A housewife would have regarded dealing with greasy fleshpots or slimy pigs' intestines as an afront to the values of her culture.[20]

The changes in household technology in the course of the 20th century brought a further aspect of clinical purity into our larders: the stereotyped whiteness of the refrigerator and deep-freeze symbolize constant cleanness. Order governs their strictly categorized compartments. The sense of smell which used to be activated in the smoked food store is now deprived of its function by the freezer compartment. In it we find foods of all kinds from all over the world, but covered in a white film of ice they lose their recognizable identities. Food must be clean.

This applies particularly to the 'ready-to-serve meals' which the technology of preserving and provision have made popular. Ready-to-serve meals not only save time in preparation; beside this functional significance they also reflect the relationship that consumers feel instinctively when dealing with food. The deep-freezing of ready-to-serve meals is the technology of food-preserving at its most civilized. The deep-frozen pizzas topped with mince or pieces of anchovy no longer remind us of the process which the pigs, cattle or fish have passed through on their way to being frozen, and we avoid the unpleasant feeling that we are connected in their slaughter and preparation.

There are no disgusting processes involved in buying ready-to-serve meals, nor in freezing them in one's own deep-freeze or heating them up. The only contact with the food in the preparation of these meals occurs in the heating and the usually hasty consumption. The neutrality of meals determined by modern technology speeds up their consumption ('fast food'). This eating is not a sensual pleasure but the fast intake of civilized fare.[21]

Shopping gives stockpiling a new face. But the standards laid down for this activity are determined by values similar to those underlying the preserving of food before the technological revolution. One of these standards is the idea of thrift. In the housekeeping guide mentioned above, (entitled *Der programmierte Haushalt [Programmed Houskeeping]*) we are told to:

> limit daily shopping as much as possible; instead make a big shopping expedition once a week and buy all non-perishable food for a month in advance. ...[Bulk buying] is quite a considerable strain on the housekeeping budget at the beginning of the month but it reduces the spending over the whole month. So you must be careful in your calculations, and above all be disciplined when out shopping or else you will return home to find things in your shopping basket which you don't really need.[22]

Or this excerpt from a 1982 handbook:

Never go shopping on an empty stomach. You will be only too easily led astray by the hunger to buy things that you don't need at all.[23]

As well as the strictly economic principles that lie behind this advice what is particularly striking is that considerable self-discipline is demanded of the shopper - and this means as a rule the female shopper. The seductiveness of things not on the shopping list must be withstood. The planned storing of provisions demands ascetic behaviour which leaves no room for a creative list of things one would like to have.

Understandably the advice given in the housekeeping text books is often modified in reality. Recommendations are only given if they are disregarded. They are the expression of a norm—'things as they ought to be'. Since the 1970s supermarkets and discount chains have had a significant influence on the provision of food in West Germany. In particular, weekly and monthly supplies of food are obtained from these large supermarkets. They have two sales approaches: first, they offer a range of goods which far exceeds that in a food shop; and secondly, they try to break down the self-discipline of the hordes of shoppers. The customer is enticed by special offers, discount and bulk prices, and super cut-price offers. Neuloh has shown that most consumers only decide what they want after they have registered the special offers.[24] The hypermarkets have names that suggest financial advantages: 'Schnell hin', 'Konsum', 'Hit-Markt', 'Depot' and suchlike[25] are examples of these supermarkets in which bourgeois thrift and virtuousness are undermined. Nevertheless the customers appear to stick to their economic principles: they buy a lot at bargain prices. The variety of the provisions unsystematically stored in their cellars reveals the hamster mentality of the supermarket customers. The presence of the large packets of ice cream, the packs of ten bars of chocolate and the family assortment of fish fingers, all bought in the whim of the moment, can be explained in terms of price, or perhaps of taste, but it is hardly ever related to seasonal factors.

In this respect the term 'hamster' ['hoarder'] is misleading—and at the same time significant. Its German usage does not refer to the foresighted, seasonal stockpiling—as seen in the behaviour of the real hamster - but to the unco-ordinated piling up of provisions[26] that the housekeeping manuals warn against. This erroneous meaning thus symbolizes the internalizing of independence from seasonal fluctuations.

It is interesting that provisions bought at a supermarket often lose some of the prestige previously associated with preserving food at home. Among large sections of the population, for instance, it is frowned on to be seen in public bearing a plastic carrier bag from the 'Aldi' food chain. Plastic bags of this sort are sometimes maliciously and dismissively referred to as 'Turkenkoffer' ['Turks' suitcases'], and to carry them symbolizes a hamster mentality and the consumption of cheap products.

People who do not preserve food themselves and prefer to buy provisions at the supermarket are catered for by the hypermarkets with their anonymous character. As a rule these are situated in the outlying districts of towns, since they do not fit in with the modern 'shopping experience' provided in the renovated inner cities. Supermarkets and discount stores are boring - drearily monotonous inside and out.

One reaction to the levelling effects of industrial food production and preserving techniques may be the revitalizing of home preserving in the last ten years or so. Every autumn the magazines for the 'modern woman of today', such as *Brigitte, Freundin* or *Für Sie*, publish 'Grandma's tips on making jam, pickling cucumbers and smoking ham'. Is it commendable to do it yourself, but reprehensible to hoard?

CONCLUSION
My paper is intended to demonstrate the significance of the way stockpiling is organized in various cultural milieus. These are only a few superficial examples and they are open to further discussion, but it should be clear that the preserving, storing and purchasing of food are not phenomena which can be just analysed simply on their own. Changes in the organization of stockpiling are not just the consequence of new technological and economic conditions; underlying the form taken by this activity we find values which were present before the revolution in household technology, and which have their origin in the structure of bourgeois patterns of culture. We can see that one of the causes of new techniques in housekeeping and preserving food is a change in cultural values. Besides the external economic, technological and social influences, individual and collective experiences play an important part: the significance of the organization of storing food should also be understood in conjunction with ideas of prestige, representation, distinction and civilization. Even such natural expressions of culture as the provision of food are the result of people creating their own worlds of ideas in differing social evironments, and their patterns of values and thought are responsible for a great variety of habits and peculiarities.

NOTES
1. *Sydsvenska Dagbladet,* 1 December 1984.
2. Wiegelmann, G., 'Was ist der spezielle Aspekt ethnologischer Nahrungsforschung?', *Ethnologia Scandinavica,* 1971, p 7.
3. Wiegelmann, G., 'Volkskundliche Studien zum Wandel der Speisen und Mahlzeiten', Teuteberg, H.J. and Wiegelmann, G.,*Der Wandel der Nahrungsgewohnheiten unter dem Einfluss der Industrialisierung* Göttingen, 1972, p 314.
4. Wiegelmann, G., *Alltags- und Festspeisen,* Marburg, 1967, pp 64 ff.
5. Kulischer, J. quoted in Egner, E.,*Entwicklungsphasen der Hauswirtschaft,* Göttingen, 1964, p 20.
6. Meiners, C., *Briefe über die Schweiz,* part 3, Berlin, 1790. Quoted in Egner 1964 (note 5), p 35.
7. Egner 1964 (note 5), p 36.
8. Egner 1964 (note 5), p 40.

9. Lewald, F., *Meine Lebensgeschichte*, Berlin, 1861. Quoted in Glaser, H., *Von der Kultur der Leute*, 1983, pp 175 f.

10. Lewald, F. 1861 (note 9), p 176.

11. Lewald, F. 1861 (note 9), p 177.

12. Cf. Teuteberg, H.J., 'Die tagliche Kost unter dem Einfluss der Industrialisierung', Teuteberg, H.J. and Wiegelmann, G., *Unsere tägliche Kost*, Münster, 1986, p. 357, and Wiegelmann 1972 (note 3), p 314.

13. Teuteberg, H.J. 1986 (note 12), p 356.

14. Wiegelmann, G. 1967 (note 4), p 68.

15. Cf. for example, Neuloh, O. and Teuteberg, H.J., *Ernährungsfehlverhalten im Wohlstand*, Paderborn, 1979, pp 140 ff.

16. Wilhelm Kieselbach, quoted in Egner 1964 (note 5), pp 41 f.

17. Frykman, J. and Löfgren, O., *Den kultiverade manniskan*, Lund, 1969, pp 111 ff.

18. Höpfl, U., *Der programmierte Haushalt*, vol 4, *Haushaltsmanagement*, Munich, 1971, p 97.

19. Elias, N., *Über den Prozess der Zivilisation*, Frankfurt am Main, 1976.

20. Frykman, J. and Löfgren, O. 1979 (note 17), pp 138 f.

21. The opposite development leads to an intensified ritualization of eating, which is related to aestheticism and luxury consumption. See Rath, K.D., *Reste der Tafelrunde*, Reinbeck, 1984, pp 280 ff.

22. Höpfl, U., 1971 (note 18), p 102.

23. *Tausend tolle Tips fur Haushalt und Familie*, Deutscher Bücherbund. Stuttgart and Munich, 1982, p 28.

24. Neuloh, O. and Teuteberg, H.J.,1979 (note 15), p 133.

25. These names are from the city of Münster and its environs.

26. *Der grosse Duden. Rechtschreibung*, Mannheim and Zurich, 1967, p 315.

A CONTEMPORARY LOOK AT OLD FOOD PRESERVING METHODS PRACTISED IN MORAVIA

Miroslava Ludvíková

In the course of time some food preserving practices die out, to be replaced by new, usually less laborious, ones. However, preference is clearly given to traditional methods of production and taste.

In an extensive, subsistence-orientated economy, with only surplus produce sold on the market, people had to make allowances for seasonal variations in supplies and ensure that there was also sufficient food on hand to carry them through the winter and early spring. On the larger farms, part of the labourer's wages took the form of board. Thus, round about 1900, it was often only the additional supply of good food that made it possible to compete with the higher wages paid to a factory worker. Grain, flour, potatoes, pulses and, in some cases apples or eggs, merely required suitable storage. Other foods had to be treated differently. Under the circumstances, the methods used included fermentation, as well as the removal of excess water by drying, curing and thickening.[1]

The foods of animal origin preserved were dairy produce and pork. In the case of dairy produce, this led to the production of *Olmützer Quargel*, which are classed among the 'sour cheeses'. To make it, the skimmed cow's milk was allowed to sour and a piece of bread crust placed in the milk to accelerate the fermentation process. The sour milk was then slowly warmed, without any additives, until it coagulated to form curd cheese. The excess whey was then separated off in a press. The mass of curd cheese was salted, kneaded thoroughly and formed by hand into little pellets which were then slapped hard against a pastry-board to flatten them into discs, after which they were dried in the sun. At this stage, they were highly suitable for grating. To make *Quargel*, on the other hand, the discs of curd cheese were placed in wooden containers to mature. For this purpose, they were moistened with salt water, beer or wine, thus promoting the growth of micro-organisms that decomposed the proteins, giving the products their distinctive taste and characteristic aroma. It was always the consumer who decided on how ripe they should be.

Formerly, they were eaten with bread and usually washed down with alcoholic drinks, particularly when there were many guests to be catered for, as, for example, in the Hana Plain when people were welcomed to a village, or at a traditional funeral feast, but most often during snack breaks at work.[2]

In the Olomouc region, there was already evidence of *Quargel* production for the market at the turn of the 15th and 16th centuries. But it was only the penetration of the Hana Plain by capitalist production methods, the swift development of the railway network and the creation of dairy co-operatives that brought an upturn in production that lasted up to the outbreak of the First World War.[3] This business, which long owed its prosperity to the dexterity of the 'slapping women', eventually ousted the domestic production of *Quargel* completely.

Before the First World War, *Olmützer Quargel* were exported to all the 'laender' in the Austro-Hungarian Empire, as well as to the Balkans and the West. After the collapse of the Empire, only the newly founded enterprises in Austria (e.g. in Tulln) were successful outside Czechoslovakia. Even in Moravia itself, production gradually declined on account of the severe competition from soft processed cheese.[4] And yet *Quargel*, with their high protein and low fat contents, in fact constitute a food of great nutritive value that meets the requirements of the modern diet. They still enjoy great popularity today (even if less than formerly) on account of their sharp, piquant taste. They are very useful in the preparation of other dishes.

Figure 1. The oldest existing photograph of Quargel *making. Prostojev, about 1880 (from Pospě, 1986).*

Smoke-drying is another method of preserving. This is the only known way of preserving pork and bacon once the pig has been killed.[5] The meat and flitches of bacon are cut into longitudinal strips, salted down, rubbed with garlic and then pressed into a tub. The tub is then put in a cool place and its contents are turned over after a few days. After 2 to 3 weeks, smoking can begin.[6] *Knackwurst* of coarse-cut, seasoned meat did not require any prior storage. Formerly, the meat was smoked in the chimney, or in special smoking houses. People learned from experience which wood was most suitable for smoking. The wrong wood could, in fact, ruin everything.[7] The pork and bacon rinds were hung in the attic and, apart from meeting everyday needs, were used in the special dishes served on Shrove Tuesday and at Easter. They were offered to guests or people in some official capacity, generally to men.[8] *Mährisches rauchfleisch* or Moravian smoke-dried meat and *knackwurst* are well-known commercial specialities.

Meat curing was solely men's work. And so it has remained down to this day. Despite the break in continuity in the rural areas, the smoke-drying of salted meat was also taken up by men from the towns who also lived there, either in suburban villas or as the owners of weekend houses. Here, profession plays no part. The passion for this type of curing grips scientist and workman alike. The new smoking chambers are made of wood or tiles; the pork, of course, is bought on the market.

Among preserved foods of vegetable origin, the cabbage holds pride of place. In Europe and Asia, it is one of the commonest types of vegetable (it was cultivated more frequently in the Bohemian regions than in Slovakia).[9] From time immemorial, it has been preserved by lactic fermentation. According to written evidence dating from the 17th century, in both Southern Moravia and Lower Austria, it was stored in deep, board-clad pits after being cut with the cabbage slicer and then pre-cooked.[10] Our research, as well as the relevant regional literature, tells us only about the storage of raw cabbage, which was cut with the slicer and pressed into barrels layer upon layer, with the addition of salt, cumin, onions and apples. In Eastern Moravia, to accelerate the fermentation process, it was common to add a piece of bread crust or a portion of sourdough wrapped in a cabbage leaf.[11] The filled tubs were kept for 10-12 days in the living room, after which they were placed in the pantry.

Marinating the cabbage was an activity in which all the family took part. Sauerkraut, far from being a supplementary food, formed, together with turnips, the nutritional basis for the entire winter, even long after the introduction of potatoes.It is still popular to this day, although consumption is noticeably declining. It is still one of the typical preferred foods for festive occasions. After a short interruption, sauerkraut marination today enjoys renewed popularity, both in rural communities and in the suburban villa or weekend house. Commerce has responded to this trend by offering special earthenware recipients of suitable sizes.

Another vegetable, which was not grown before the 16th century and the preserving of which became a South Moravian speciality, was the cucumber. It was first marketed pickled in brine with dill. Later, they were placed in two to five-litre jars together with vine and mahaleb cherry leaves, as well as mustard seeds, and soaked in brine and a 10% vinegar solution (in a proportion of 1:1). Although these preserved cucumbers were very sour, there was a ready market for them before the First World War. And so there was no surplus for export outside the Empire. Around 1930 it was discovered that natural sugar, which was not used because it was assumed to cause unwanted fermentation, could be successfully replaced by a synthetic sweetener.[12] It was then that the sour-sweet *Znaimer Gurken* became a by-word for quality, and their export increased substantially. Today, large quantities of them are preserved using commercially available pickling salt, even in households that do not possess their own kitchen gardens.

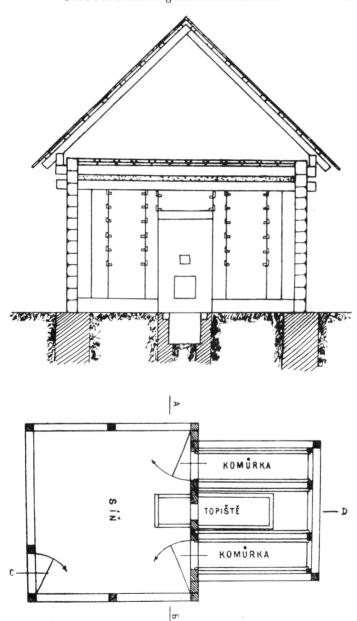

Figure 2. A fruit drying chamber from the Moravian Wallachia in the 19th century (from Horák, 1893).
   komurka - chamber
   topiste - oven
   sin - vestibule

Before turning to the last group of preserved foods, we should mention the methods used to preserve mushrooms, which cannot be classed as either a vegetable or a fruit, but which has become a very popular pastime to pick. Today, as in the past, they are preserved by drying and marinating.[13]

Fruit preserving made an important contribution to the food supply and has continued to do so, even if the methods have changed in the course of time. But fruit itself represented an essential addition to one's diet, in every form, whether fresh, dried or stewed, i.e. boiled until nearly all the water has evaporated. The result is a thick paste, sweetened by the concentrated fruit sugar. Fruit growing in Moravia was spread by the monasteries. It was not until about the middle of the 18th century, however, that it emerged from a period of relative stagnation due to three hundred years of continual war. The most widespread types of fruit, particularly in East Moravia, were the different varieties of plum. For instance, 200 quintals of dried plums from the Uherské Hradiště region were offered for sale on the Prague Ungelt as early as 1597. Dried fruit (and nuts) were exported in large quantities from South Moravia far out over the frontiers, providing the country with a major source of income. From North Moravia, waggoners brought this produce as far as Austrian and Prussian Silesia.[14] The occupation of Bosnia and Herzegovina by Austria-Hungary, of course, led to fierce competition in this line of trade.

In small households, the fruit was dried on hurdles in the oven after the bread had been baked. Slices of apple were threaded onto strings and dried on the parlour stove. The fruit was often dried on a large scale in purpose-built drying houses which were a particularly characteristic feature of East Moravia. In the mountains, these were built of stone blocks and, in the lowlands, of loam and straw. They usually contained two rooms, sometimes with the fire door and the flue beneath an extended roof. The larger room was used for preparing the fruit. As the operation represented 6 weeks' uninterrupted work, on behalf of several parties, the stoker also had his bed here. The oven stood in the middle of the wall. The smaller room around the oven was surrounded by large drying hurdles which could be pushed onto gratings and enclosed by flaps.[15]

The greatest care had to be taken when drying plums, but apples needed less attention. Pears were dried whole.[16] If the plums were fleshy and smooth, and there were about 90-100 of them to the kilogramme, they were able to compete with the produce of Bosnia. They were used in a great variety of ways in the home—in addition to the customary desserts, plums and dried pears were served as an accompaniment to the traditional winter dishes. They were the only delicacies enjoyed by the children, who were given them on the occasion of the Christmas and spring processions.[17] In South Moravia, people would suck dried apricots for refreshment as they worked in the fields on a hot summer's day; dried cherries were used as a substitute for raisins.

Nowadays, fruit is dried only in exceptional cases, for the traditional Christmas dishes, for instance, but also for normal consumption without any further preparation. Plum drying in East Moravia was ousted by *slivowitz* distilling. Windfalls are now turned into fruit wine in the presses of the allotment gardener associations. Large volumes of fruit are stewed at home to make compotes, although the market is well stocked with such products.

Plums used to be stewed to make a thick paste; this activity took place in the autumn and still brings back fond memories as it gave the young folk an opportunity for socialising and entertainment.[18] Originally, the paste was cooked in an earthenware cauldron sunk into the garden beneath a temporary awning. As it was not possible to remove the stones of *Duranzen*, a variety of plum, from the flesh of the fruit when it was raw, it was first cooked to a pulp and then pressed through a deep earthenware sieve using a plunger. After many hours of thickening, the plum mash was ready to be placed in vats, after which its surface was strewn with dried elder blossoms. In this way, it would keep for at least three years.

While fruit drying was the business of the men because of the hard work involved in restacking the drying hurdles, a number of older women were engaged to make the thick paste. This long drawn-out task would have been too much for one woman on her own, particularly if the paste had to be made for several different parties. Later on, it became the practice to install the cauldron permanently in an oven of unburned tiles in the shed, the earthenware cauldron being replaced at the same time by a copper one equipped with a mechanical stirrer.

Plum paste also served as payment in kind, as well as being exported like the dried plums. After the price of sugar dropped, it was supplanted in the towns by marmelade, which was less dehydrated. Its place was later taken by various kinds of jam. In rural areas, this transition was complete by the 1950s. At the present time, however, particularly among diabetics, efforts are being made to bring back the paste making as no sugar is needed. At the same time, new techniques are being sought after, of course, as the traditional methods, which require considerable time and attention, are no longer in keeping with the spirit of the modern age. Another drawback is the risk of crop failure due to bad weather.

To sum up, we can say that, as in all other manifestations of folk cultures, some practices die out in the course of time (if not completely, as with traditional fruit preserving methods), to be replaced by new methods which, although they may be less laborious, are not always above reproach from the health viewpoint as, for instance, when they use excessive amounts of sugar. Some traditional techniques disappeared for a short time and have now been revived. The initial enthusiasm for the great god progress that came with the emphasis on large-scale, mass production and the attendant reduction in hard manual labour was soon to wane. Today, sauerkraut making and pork curing are part of domestic activities. In the latter case, conservation is not, in any case, the main concern, since the raw material is bought on the market and the smoked meat is consumed shortly after curing. Preference is clearly being given here to traditional values in matters of production and taste that are not very amenable to

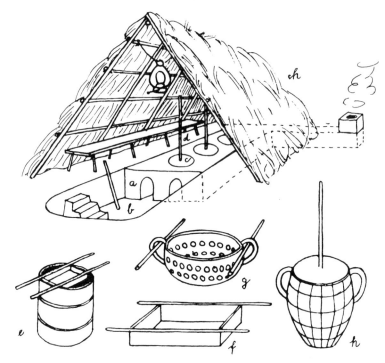

Figure 3. Traditional method of cooking thick plum paste around 1900, as illustrated in an old village information bulletin from Nemcany near Slavkov.

ch - hut
a - oven
b - pit for wood
c - built-in caldron (= h)
d - stirring stick
e - tub with strainer for stewed plums
f - wooden strainer with iron sieve
g - earthenware strainer
h - wired earthenware cauldron with stirring stick (= c,d)

change. Men, in particular, take the view that the use of chemical preparations prior to curing (to control the curing process) substantially impairs the taste of the end product.[19]   As to cheese consumption, *Olmützer Quargel* are still highly popular, despite the other types of cheese on offer, particularly among the large population of beer drinkers.

The present trend towards domestic food preserving is no longer, therefore, prompted by necessity but by thrift. For instance, people either preserve their own excess fruit and vegetable produce or go to market, if it is cheaper to do so. The raw meat that we buy is also cheaper than cured meat.   And then there is the matter of taking pride in one's own achievements. ('I made that myself'), i.e. reasons of prestige: women pride themselves on their fruit and vegetable preserves, men on their cured meats. A third reason is probably the distrust of modern mass production methods, above all the use of chemical additives, which are not present in home produce. At the same time, self-support enthusiasts are well aware that carcinogenic substances may be formed during curing, for example, whether it be at home or in large industrial plants.

We must also take into account the way in which preserving has been affected here by the advent of freezers. Out in the country, they are even supplanting the modern method of preserving meat in jars or tins. They have made less impact upon urban households as no seasonal or market-conditioned price advantages come into play here. The freezer has not ousted curing. The preserving of cucumbers and vegetable specialities also remains a domestic activity. The varieties of deep-frozen vegetable that are to be found on the market in the cities also set the pattern for the rural areas. As to fruit, only strawberries and plums are placed in the freezer; other species are bottled.

Finally, those products that represent the Moravian art of food preserving abroad deserve a brief mention, above all *Znaimer Gurken* which are served, for example, at modest promotional receptions, along with Moravian cured meat, Moravian *knackwurst* and *slivowitz*. In this combination, they have come to symbolise the old country for all emigrants, especially those overseas.

NOTES

1. Nowadays, the daily newspapers also give tips on short-term food conservation. For instance, if the refrigerator is out of action because of a power cut, it is recommended to wrap meat in horseradish leaves (guaranteed conservation: 24 hours) or in nettles (guaranteed conservation: up to 2 days). Here, it is the formic acid that acts as a preservative. Farm butter used to be taken to market packed in these leaves; out in the country, only butterfat was consumed.

2. As from the 16th century, goat's cheese and ewe's milk cheese were also made in the mountain dairy farms of Moravia.

3. The origins and development of this product were illustrated at an exibition held in the Olomouc museum in 1986. See Pospěch 1986, 9f (*et seq*) (the photograph was also taken from this exhibition).

4. The methods of producing this type of cheese were also known in the folk culture, if not everywhere: the curds were mixed with salt and cumin and then left to rest for three days before being heated up with butter.

5. Formerly, when the pig had been slaughtered, its head was also cured; nowadays, the meat is put into liver sausage. In addition to meat, chunks of cottage cheese (Jancár 1966, 56) and ewe's milk cheese were smoked in SE Moravia, and even turnips in the poorer areas.

6. For details, see Stika 1980, p 83.

7. Hardwood has to be used, mainly the wood of the plum tree in the lowlands, occasionally that of the pear tree, and, in the mountainous areas, beech or hornbeam. Oak is unsuitable (because of the sour aftertaste it gives the meat).

Spruce should be avoided because of its terpentine taste, while willow gives off too much smoke.

8. See Václavik 1930, p 136.

9. On a worldwide scale, cabbage is cultivated on approximately 10% of the surface devoted to vegetables. Comparable areas are cultivated in Czechoslovakia, East Germany, West Germany and Holland (Bartõs 1986, p 505).

10. For details on the storing of cabbage in pits in Austria, see Schmidt 1966, pp 336-7. In Moravia, this technique was used on the manorial farms. In Vrakov in 1604, for instance, one of the tenant's duties was to steam the cabbage, which was then cooked for one day and placed in the pits or in vats (Hurt 1969, p 30).

11. See Falesniková, 1926, p 61, and Václavik, 1930, p 136.

12. See Vilikovský, 1936, pp 542, 551, 558.

13. Dried mushrooms were also exported from SE Moravia to Vienna (Václavik, 1930, p 132).

14. See Vilikovský, 1936, pp 62 and 68; Cedidla, 1966, pp 50 and 53.

15. Descriptions of these constructions and articles on fruit drying abound in technical literature (see Horák, 1893, pp 24-6; Trnka, 1926, pp 56-8; Vicenik, 1944, p 13; Zila, 1947, pp 48, 21 *et seq*, etc; see also the illustration).

16. Ground dried pears (carrots too on the Drahan Plateau) were still used for sprinkling in the 1920s.

17. On St. Nicholas' Day, for instance, the children in the Bojkovice neighbourhood received as presents dried plums with walnuts in place of the kernels stuck on boxwood rods (Vicenik, 1944, p 13).

18. Considerable literature has been devoted to cooking plum jam (Sebestová, 1947, p 117; Vicenik, 1945, p 10; Cedidla, 1951, pp 184-5; very detailed description in Rysová, 1963, etc; see also the illustration).

19. Control of the curing process by adding chemicals is nothing new. As early as 1825, M.D. Rettigova recommended adding one pound (512g) of salt to one lot or half-ounce (16g) of saltpetre; the meat could then be cured after 8 days, the saltpetre together with the salt prohibiting unwanted bacterial growth.

"MARINERS' MEALTIMES": The Introduction of Tinned Food into the Diet of the Royal Navy

Una A. Robertson

It is not surprising that the basic provisions supplied to the Royal Navy during the 18th century consisted of foodstuffs of a long lasting nature. Ships might well be at sea for anything up to six months without revictualling. So items such as beer, salt meats, pease, hard biscuit, oatmeal and vinegar, had long been seen as suitable provisions for seafarers.

Although the method of preserving food in glass jars had been discovered in France and publicized in 1810, it was in England that the idea was taken up and further developed. These 'preserved foods' were introduced by the Navy Board with surprising speed, but initially on a small scale and only as medical comforts for the sick. After some years, however, tinned foods were to replace some of the salt meats and other items.

Finally, consideration is given to the question whether these tinned foodstuffs made a significant contribution to the sailor's diet and how soon such an improvement was apparent.

In April 1814 the *Edinburgh Review*, a periodical then at the height of its circulation, carried a lengthy article on the subject of food preservation in general and the work of M. Appert in particular. In discussing the Frenchman's method of preserving foodstuffs in glass bottles, the article recognized that:

> By such means, not only may the more perishable alimentary substances of one season be reserved for consumption at another, but the superfluous productions of distant countries be transported to others, where they are more needed. To mariners, in particular, every means of preserving articles of subsistence in a recent state, must present an object of great interest; and even though this should not be practicable to the extent of supplying daily food for a large crew, yet an occasional use of such food would be at all times a great luxury and, in many cases of sickness and disease, essential, perhaps, to the restoration of health.

Even at this early date the *Edinburgh Review* appears to have appreciated not only the significance of the invention and its possible benefits, in which particular recognition is made of the needs of mariners; but also the link between diet and health, for so long unrecognized or ignored. Even more remarkably, the article accurately describes the situation in the British Navy over the next forty years or so.

To understand any potential significance these new preserved foods may have had for the British sailor, we must begin by looking at his diet in the years prior to M. Appert's invention.

From the time their travels had taken them out of reach of land for long periods there had of necessity been an element of preservation about the foods considered suitable for seafarers. Voyages were often of many months duration, with no certainty of fresh supplies being available on the way. Add to this the problems of providing for the very large numbers of men required to work the ship, plus the difficulties of storage in permanently damp mobile conditions, and it is perhaps not so surprising that there should have been an underlying unanimity about provisions taken on board. For centuries the predominant ingredients of a seaman's diet had been salt meat, salt fish, dried pease and beans, biscuit and beer. It made little difference whether the ship was of a mercantile nature or sailing on behalf of King and Country. The diet was recognizably the same. The seaman employed by the East India Company in the early 17th century would have been familiar with the diet of his Elizabethan cousin some fifty years before: equally he would have felt at home on board one of Nelson's ships some two centuries later. In this instance, however, there would have been one notable exception, although doubtless he would have taken to it as readily as the proverbial duck takes to water.

Throughout this period the powers-that-be laid down a scale of provisions for each man on either a daily or weekly basis, and there had been few changes over the years. By Nelson's day the scale had crystallised into 1 lb biscuit and 1 gallon of beer per man per day: and on a weekly basis 4 lb salt beef, 2 lb salt pork, together with some oatmeal, pease, cheese, and a little butter. There was also the daily allowance of rum. A little variation was achieved by withholding items of the basic allowance and issuing other things in their place. In foreign parts there were certain recognized substitutes that could be given out in due proportion, such as rice instead of oatmeal, olive oil instead of butter, wine instead of beer, and so on. A few innovations had crept in over the years, such as vinegar, sugar, tea and cocoa. These items had originally been taken on board as 'medical comforts for the sick', a phrase of which we will hear more shortly; then used as occasional substitutes, before becoming part of the regular rations.

These schedules were only brought into use once the fresh provisions had been exhausted. While in harbour the men were theoretically fed on fresh foods, and it is clear from contemporary records that efforts were made to supply these for as long as possible after sailing. When in port, ships would be surrounded with bumboats selling local produce; and memoirs often recount how fresh fruits of every description would be slung from the hammock nettings after a stop at Madeira, the West Indies or the Cape of Good Hope. There were even instances of local fishing boats being forcibly stopped and raided for the sake of the fresh provisions they might be carrying. Fishing itself provided not only a welcome change of diet but also sport and amusement for those participating, and live turtles were often taken on board and stowed between the guns on the Main Deck until needed.

Many ships apparently set sail looking more like Noah's Arks than men-of-war, and derogatory remarks were made about the East India Company's floating farmyards. Assorted poultry were kept in coops as a source of meat and eggs; cows and goats were taken on board to provide fresh milk - indeed, the world wide distribution of Jersey cows is attributed to this practice; while cattle, sheep and pigs were transported 'on the hoof'. Many memoirs recount incidents concerning this livestock while at sea, and the sailors often made pets of them, pleading for the life of their special favourite when the time came for it to meet its end.

Much of this livestock was destined to benefit the Captain; either transported (at His Majesty's expense) to be sold on arrival for the Captain's private profit, or else to be a supplement for the Captain's dinner table. Although everyone on board was entitled to their official daily rations, it was assumed that anyone who could do so would augment them in whatever way they could. The ordinary seaman was rarely able to do very much. For him storage space was minimal and catering arrangements primitive. The officers in the Wardroom generally appointed one of their number to be 'mess caterer' for the duration of the voyage; and he, with everyone contributing equally to the cost, would buy in appropriate extras. The Captain while at sea lived, literally, on a different level to the rest of the ship, with his own living accommodation, servants, cook, and so on. The stories about morose, tyrannical captains of frugal habits, who declined to do more than was strictly necessary in the way of hospitality, can be offset by many others whose liberality towards their officers and generous treatment of their men, particularly those in the sick bay, earned them the respect and gratitude of everybody.

The Captain might have lived almost as well on board as on dry land, and the officers might have fared not too badly; but the ordinary sailors, for as long as they had been going to sea on a diet of salt meat, biscuit and beer, had been complaining about both quality and quantity. Warship or merchant ship, it made little difference, nor whether it was the 17th century or the 19th century—the gist of the complaints was unmistakably and depressingly the same.

'Unwholesome and stinking victuals, whereby many of them (i.e. the sailors) are become sick and unserviceable, and many are dead'[1] Beef so tainted that 'the scent all over the ship is enough to breed contagion',[2] or so hard that it resembled mahogany or rosewood and could be carved into ornamental boxes. Cheese made into jacket buttons or used for repairs to the ship's mast 'which have stood the weather equally with any timber'.[3] Oatmeal which, one surgeon said, it was cruel to expect the men to eat; and butter usually given over to the boatswain for greasing the ship's rigging.[4] Beer like the sewage of London which could only be drunk 'by stopping their breath by holding of their noses',[5] and water which appeared to be 'the extract of the ditches around Sheerness'.[6] But it was probably the ships' biscuits, the famous (or infamous) 'hard tack' or 'pusser's nuts' as they were nicknamed, turned out by the Dockyard bakeries at the rate of 70 x 4 oz biscuits every minute, their centres compressed into an incredible hardness, that received the greatest vilification. Everyone had stories to tell about them: 'Biscuit ... so light that when you tapped it on the table, it fell almost into dust' said one,[7] the reference being to the permanent infestation of maggots and weevils. Sailors ate them after dark 'when the eye saw not, and the tender heart was spared', said a second,[8] while a third commentary ran: 'Bread, it is well remarked, is the staff of life; but it is not quite so pleasant to find it life itself and to have the powers of locomotion.'[9]

If the complaints about quality were eloquent, the men were equally vociferous about the quantities. Although it was laid down in a series of regulations exactly how much of each item was to be supplied, it is quite clear that the men were often defrauded, by all manner of means, of the seemingly generous allowances and had legitimate cause for complaint. A résumé of maritime affairs first published in 1685 took the form of dialogues between an Admiral and a Captain; in one the Admiral is made to ask: 'But how comes there to be any fault in this kind, for as touching the quantity of the victual, I have heard it generally and confidently spoken that there is no Prince or State that maketh so large an allowance of victuals to seamen, as his Majesty doth? Whence is it therefore that there hath been such complaining of late in this kind?'[10] Over the years there were other confident assertions of this nature. The seaman however knew differently.

Their Lordships of the Admiralty could specify quantity and quality until the end of time but the entire system of victualling was riddled with malpractices. Contractors knew that they could easily bribe an official to pass as fit anything they chose to put in the casks; the purser, the man on board ship responsible for the storage and distribution of food and drink, was exposed to enormous temptations in the course of his job and had unlimited opportunities to defraud the sailor. Even the Captain might be involved; either by allowing himself to be bribed to turn a blind eye, or by actually permitting the malpractices in return for a share of the profits. Every time it would be the sailors, on the sharp end of these tricks, who were the sufferers.

The dangers of allowing the seamen to go naked or hungry had often been spoken of and over the years words of warning had been addressed to the Admiralty by compassionate commanders. One such dates from 1629. The writer pleads: 'I beseech your Lordships for the honour of the state suffer not the service to become a scandal, but be pleased to take speedy course for the redress hereof; for foul winter weather, naked backs and empty bellies make the common men voice the King's service worse than a galley slavery: and necessitous wants together with famine pleading the cause of their disorders, lays open a way to what they are too prone already to - mutinous disobedience and contempt of all commands, for necessity hath no law.'[11] These words and others similar must have echoed in their ears some 170 years later when the Admiralty was faced with the Nore and Spithead mutinies, in 1797. No mention was made by the mutineers of the expected grievances such as the brutality of the captains or the iniquities of impressment. Significantly, the sailors asked for better pay (unchanged for *145* years!), better food (more particularly a supply of fresh vegetables), and the full weight of rations allowed to them.

After such a catalogue of economies and neglect on the part of almost everyone in authority the time was surely ripe for improvement; and one might think that any contribution to the health and welfare of the Navy, so important in the defence of the realm, would have been eagerly anticipated and actively promoted. Let us see what happened over one important such contribution.

If the Admiralty faced perennial problems in provisioning the British Navy, so Napoleon had similar difficulties in feeding his vast armies. In the circumstances perhaps he really did declare, 'An army marches on its stomach'! By the end of the 18th century, French agriculture and industry were at a low ebb and prizes were offered by the Society for the Encouragement of Industry for possible solutions to their problems. M. Appert was one of the first to be awarded a prize for his invention of preserving foodstuffs in glass bottles. He had begun by experimenting with champagne bottles. When filled with different foods, their mouths were corked up and the bottles immersed in a water bath; they were then heated for varying times depending on the contents. A small factory was set up south of Paris and the products were sold at a shop in the city. They soon came to official attention. The French Navy, which carried out trials in 1804 and again, more rigorously, in 1806, reported most favourably, and M. Appert was awarded 12,000 francs on condition he publicized his methods. In February 1809 the *Courier de l'Europe* waxed lyrical: 'M. Appert has discovered the art of fixing the seasons. With him spring, summer and autumn exist in bottles like delicate plants that are protected by the gardener under a dome of glass against the intemperance of the seasons.'

The process devised by the French was taken up enthusiastically by their enemies across the Channel, where similar experiments had been underway. In 1807 the London Society of Arts had awarded a prize to Thomas Saddington for preserving fruits 'without sugar for house or sea stores' (that recognition again of the special needs of sailors). Saddington limited himself to fruits, whereas Appert by this date was successfully producing a range of products including meat stews, soups and milk, as well as numerous fruits and vegetables.

Appert's book appeared in June 1810. An English patent was filed that year describing an almost identical process using 'vessels of glass' but adding the significant alternatives of 'pottery, tin, or other metals, or fit materials'.[12] However, it was to be the engineering firm of Donkin, Hall and Gamble which initiated production of these preserved foods in metal containers. (They claimed to have paid £1,000 for Appert's patent but no trace of such a transaction was found in the records of Chevalier-Appert, the firm set up with that 12,000 franc prize money.) By August 1813, after many set-backs, they were in a position to offer their preserved meat for trial in some of His Majesty's ships, for the use of 'the Sick and

Convalescent'—that catchphrase again. Quite amazingly, and totally out of character, the Admiralty agreed to their proposal.

Now why should an organization that had shown itself to be opposed to change for so long suddenly show a constructive interest in such an innovation? It would be nice to think the sailors' welfare was at last being considered. Maybe it was. But other factors must also have stimulated their interest.

The complaints of the men have already been touched on. Twenty years of war and the continuing problems associated with provisioning such vast numbers must have made some impact. Contemporary estimates suggested that there were some 25,000 men in the Navy in 1793: the numbers swelled to over 145,000, and even after Trafalgar had removed all fear of invasion, were not reduced until the end of the war.[13] A ship such as Nelson's 'Victory' accommodated some 900 men, while even a 5th rate frigate carried 300 or so. Naval policy required these ships to stay at sea for long periods, to maintain effective blockades against the French; to take on board supplies for six months was quite common. So it was obviously desirable to improve such supplies and thus enhance a ship's fighting potential.

Linked to this must have been the appalling death rates of the sailors during the war years. Contemporary figures estimated that 50% of their deaths were due to disease whereas under 10% were actually due to enemy action. While the fevers, typhus, lung infections, constant bowel and stomach disorders and suchlike could all be attributed to a variety of causes, scurvy was directly connected to diet. It had haunted seafarers for so long and had been responsible for untold misery and innumerable deaths. Although scurvy was more or less absent from British ships during the Napoleonic Wars, thanks to the enforcement of a daily dose of lemon juice and greater attention being paid to the diet generally, there was always fear of a recurrence. The cure had been proved beyond all doubt by the famous Scottish naval surgeon James Lind in his so-called 'Salisbury Experiment' of 1747; but for many years it had been neglected as being too simple and unscientific. Even though the cure had been found, the reasons why it worked were not understood; since tinned food was initially regarded as being as good as fresh, it was anticipated that it would be beneficial in combatting scurvy.

Whatever the reasons for the Admiralty's interest, the firm's offer was taken up and their preserved meat sent for trial both with the Channel Fleet and in the West Indies, with a third batch being sent to the East Indies. In April 1814, which by coincidence was exactly the date of the article in the *Edinburgh Review* already quoted from, Sir Alexander Cochrane, the newly-appointed Commander-in-Chief of the North American Station, suggested a consignment might be sent out to him. He wrote:

HMS Asia, Bermuda
3rd April 1814

Sir,

Being given to understand that the Patent Prepared Meats and Soups are found to answer very well even in the West Indies, ... I request you will be pleased to suggest to my Lords Commissioners of the Admiralty ....for some of it to be sent out here for the sick on board the ships of the Squadron, which might be supplied to them in lieu of their salt provisions ....Upon this quarter of the station, where there is not a possibility of getting Fresh Meat to make Soup, excepting at a most enormous rate, a supply of the Article would be a most desirable thing for the Sick. If a supply of these Soups and Meats could also be sent to the Hospital at Bermuda, it might be the saving of great expense ...[14]

The following year Mr Grimstone, surgeon of Cochrane's flagship, was enthusiastically writing to the Board: 'By an order of the Commander-in-Chief dated at Bermuda, the surgeons are directed to report to the Board their opinions as to the use of Preserved Meats on board ship. In consequence I have to state that it appears to me to be infinitely superior to anything I have seen in use ... In long cruises in the tropics, I am persuaded it would tend very much to prevent the frequent and dreadful consequence of scurvy, by being served out generally one or two days in the week.'[15] (He meant served out to the whole crew, not merely the sick). However, a few of the tins were condemned as being wholly defective and extremely offensive, thought to be due to damage in transit.

Testimonials rolled in from other sources. As the Navy adapted to peacetime conditions after 1815 it was involved in exploration, survey work, and policing the seas against piracy and slavery. Tinned foods proved their worth time and time again. A long and enthusiastic testimonial was given by the Russian explorer, von Kotzebue, who took some of these 'English provisions' to the Behring Straits in 1815, as well as Russian-dried cabbage and meat. These latter were made into a dubious stew one evening whereupon he had 'two tin boxes of English patent meat opened', he said, 'to take away the bad taste of the soup'.[16] The following year tinned foods accompanied Lord Amherst on his mission to China and were highly commended. During the 1818 Baffins Bay Expedition under John Ross the men were issued with 1 lb preserved meat and 1 lb vegetable soup each week; and in 1820 when Parry (previously 2nd Lieutenant to Ross), returned from leading his own Arctic expedition he enthused over the tinned meats and soups supplied by Donkin and Hall. The Surgeon and Assistant Surgeon added their praises. The surgeon on HM Ship 'Hecla' said: 'I consider them to have been acquisitions of the highest value ... and am also happy in testifying to the general good quality of these provisions.' His assistant wrote: 'I have no hesitation in pronouncing my opinion that to the judicious employment of these articles is to be attributed, in great measure, the preservation of the general health of the Officers and Crew.'[17]

Donkin, Hall and Gamble became established as suppliers to the Navy and in January 1817 preserved meats made their first official appearance in the Victualling Schedules, creeping in as usual under the guise of medical comforts. They replaced 'portable soup' made from those slab-like lumps of glue (derived from ox offal boiled up with vegetables), invariably rejected by all to whom it was fed. On offer the following year were fifteen versions of tinned meats and soups.[18] Soon after this surgeons were instructed that: 'The said meats are to be issued gratuitously to such of the sick and convalescent on board as may, in the Surgeon's opinion, stand in need of such refreshments, in small quantities of from 2 to 6 oz per man per diem: and these quantities are not to be exceeded unless under very peculiar circumstances.'[19]

The next significant date is 1831, when these supplies were directed to be carried as medical comforts on all ships rather than on certain categories as before. The final step in our story took another 16 years to accomplish; their transfer from the sick list, so to speak, and their adoption into the general dietary. The provision scale was changed so that on alternate salt beef days $3/4$ lb preserved meat was issued with $1/4$ lb preserved potatoes or $1/4$ lb rice. The sailors had problems in gaining access to the tins (tin openers had not yet been invented) and it was directed that a special lever knife was to be given out to each mess. Another circular warned that if, on opening the tin, there was an offensive smell the contents should be discarded.

Unfortunately this was happening increasingly often. The Navy had been unlucky in their timing. Although Donkin's firm had been the original naval suppliers, other contractors duly appeared on the scene, including one Stephen Goldner who had patented an entirely new process. In 1845 he had supplied Franklin's expedition with tinned provisions, a proportion of which had had to be condemned. He was also a large-scale supplier to the Navy. In the interests of economy he was attempting to supply much larger cans—for example, an entire sheep—and through an imperfect understanding of the process Goldner came to grief. Serious defects in many cans were reported by the Victualling Yards and this led to widespread mistrust of all tinned foods.

A Parliamentary Committee was set up to investigate 'the dates and terms of all Contracts for Preserved Meats ...with Goldner', the specifications of each contract ('meat of the best quality and cannisters of the best tin'), whether there had been any compaints, whether these meats had been issued for Arctic voyages and so on, right down to the last half-penny costing per can.[20] A second Inquiry investigating the supply of salt meats, demanded intricate details about quantities, packaging, prices, the origin of the meat and destination of the tins, and whether any had been rejected and the causes for the rejection. A specimen contract was included which said that the beef for salting should be 'prime good sound sweet fat and well-fed' and that no 'unusual pieces shall be packed with the said beef'.[21] Enough to make a sailor laugh! He reckoned he was given salt horse, particularly when he met the hooves and the harness in the barrel along with the meat. A further Inquiry in the 1860s compared the annual costs of supplying salt meats and tinned,[22] but this was after the Navy had set up its own cannery. The venture was short-lived: the sailors however seized the opportunity to invent yet another of their colourful nicknames,[23] just as they had immortalized Appert's preparation of boeuf bouilli by calling it bully beef.

Finally we have to consider the difference tinned foods made to the diet of the British sailor. It would be nice to report that they all grew hairy chests and lived to 108, but I cannot! The Navy of the 1850s was very different to that of 1815 and undoubtedly healthier, but this could be attributed to numerous causes. Wars were fewer and periods at sea shorter. The installation of iron water tanks filled by hose not only ensured a healthier water supply and lessened their dependence on alcohol to slake their thirst but also, by acting as ballast in place of the old style wooden water barrels and shingle, removed a festering source of infection. Distillation apparatus, although known about for centuries, was also introduced during this period. The greater regard for the welfare of the men saw such measures as the reduction in the rum ration (1825 and 1850) and the introduction of new provision scales (1824 and 1850), pensions after 22 years service (1831), libraries in ships at sea (1838) and rest homes on shore (1835). Better conditions and pay attracted a better calibre of sailor and the introduction of continuous naval service (1834 onwards) turned the Navy into a career rather than an interim measure.

The greater attention paid to diet has already been noted. Tinned foods were initially credited with anti-scorbutic properties and their reputation was considerably enhanced by the healthiness of the earlier Arctic expeditions. Generally speaking such voyages were provisioned with the greatest care and included other items such as lemons, cranberries and sauerkraut.[24] The noticeable improvements as regards scurvy could not really be attributed to the benefits of tinned foods. During the Napoleonic Wars there were relatively few cases on board naval ships and as early as 1803 Thomas Trotter, yet another eminent naval surgeon, and yet another Scot, wrote that he knew of no scurvy cases needing hospital treatment since 1795. The daily dose of lemon juice might have helped, but it was such a small proportion of the sailor's daily requirements that other factors must have been responsible. Interestingly, scurvy continued to plague the Merchant Service and it was in 1844 that masters were first compelled to serve crews with lime or lemon juice after ten days on salt meat.

If it is impossible for us to judge the part played by tinned foods in improving the standards of naval health and welfare, we do know that the medical men of the day thought highly of them; their statements are on record as we have seen. We know how captains and admirals regarded them because they too recorded their thoughts. One such was Basil Hall, the Scots born captain of the 10 gun brig 'Lyra' escorting Lord Amherst to China. He had entered the Navy in 1802 and was quick to appreciate the benefits of tinned provisions over meat transported the traditional way, on the hoof. He was later to write:

'you must ... bear in mind that meat thus preserved eats nothing nor drinks - it is not apt to die - does not tumble overboard or get its legs broken or its flesh worked off its bones by tumbling about the ship in bad weather - it takes no care in the keeping - it is always ready, may be eaten hot or cold, and this enables you to toss into a boat as many days cooked provisions as you require - it is not exposed to the vicissitudes of markets, nor is it scourged up to a monstrous price as at St. Helena, because there is no alternative. Besides these advantages it enables one to indulge in a number of luxuries which no care or expense could procure.[25]

And what did the men think of them? Well, many people over the years had remarked on the innate conservatism of the British sailor. Samuel Pepys, when Secretary to the Admiralty, warned about the difficulties inherent in changing the seamens' rations; James Cook the explorer noted that any and every innovation, however much to their advantage, would inevitably be resisted; while John Cochrane, another member of that distinguished naval dynasty, wrote simply: 'It is certainly a difficult thing to get the lower classes of men to alter their usual habits.'[26] On the one hand the Admiral in those maritime Dialogues referred to earlier remarked: 'Our common seamen are so besotted in their beef and pork that they had rather adventure on all the calentures (fevers) and scurvies in the world than to be weaned from their customary diet or lose the least bit of it.'[27] On the other hand the Return to the House of Commons (replying to whether any complaints had been made and if so when?) reported primly: 'Condemnation of preserved meats have occurred ever since their first introduction as an article of diet in H.M. Service. The earliest complaint of objectionable matter found in a canister, was received in office on 24 November 1849.'[28] The introduction of tinned food, it would seem, was liable to be fiercely resisted by the seaman.

He may not have welcomed it but what difference would it make to his diet? It would be readily available, and in better condition, when other supplies might have failed and it would have added some variety. Apart from anything else, it would take less time to prepare and less time to cook on the poor facilities generally provided; and it would be physically easier for him to eat than the inadequately soaked and boiled salt meats of the old days. Their nickname for these had been 'junk' - the name given to old condemned rope! It is right that a sailor should have the last word on the subject. 'Preserved Irish beef ...is very much overcooked in the preserving, as it takes the form of a conglomeration of strings when warmed for eating, so much so, that it has earned for itself the cognomen of 'clews and lashings'[29] —and that, for the uninitiated, means hammock cords.

Not much had changed, it seems.

REFERENCES

1. H.W. Hodges & E.A. Hughes, *Select Naval Documents*, Cambridge University Press, 1922: see under Kendall, 1653 (p 68).
2. M. Oppenheim, *The History of the Administration of the Royal Navy*, London & New York, 1896: see under Lindsey, 1635 (p 327).
3. Hodges & Hughes, *op cit*: see under William Thompson, An Appeal to the Public 1761 (p 125).
4. John Masefield, *Sea Life in Nelson's Time*, London, 1905, pp 147-8.
5. Hodges & Hughes, *op cit*: see under William Thompson (p 126).
6. Navy Records Society, *Five Naval Journals, 1789-1817*, London, 1951: see under the Rev. Edward Mangin, 1812 (p 13).
7. Jeffrey Baron de Raigersfeld, *The Life of a Sea Officer*, reprint of 1929, p 24.
8. Masefield, *op cit*, p 144.
9. Attributed to Captain Frederick Marryat.
10. Navy Record Society, *Boteler's Dialogues*, London, 1929, p 55.
11. Hodges & Hughes, *op cit*: see under Sir Henry Mervin, 1629 (p 49).
12. A.W. Bitting, *Appertizing, or the Art of Canning*, San Francisco, 1937: see reference to Patent 3372 of 1810, pp 23-40.
13. *Edinburgh Review*, Vol XLI (October, 1824): see 'Abolition of Impressment', pp 157-8.
14. National Library of Scotland MS 2348, No 11, p 8. He belonged to a prominent Scottish family, several of whom had distinguished naval careers.
15. *Naval Review*, Vol 27, London, 1939, p 103.
16. Otto von Kotzebue, *Voyage of Discovery into the South Sea and Behring's Straits*, London 1821 (I, 180).
17. International Tin Research Council, *Historic Tinned Foods*, London, 1939.
18. *Ibid:*, the versions were Mess Beef, Corned Round of Beef, Roasted Beef, Seasoned Beef, Boiled Beef; Boiled Mutton, Seasoned Mutton, Mutton with Vegetables; Boiled Veal, Roasted Veal, Veal and Vegetables; Soup and Bouilli, Vegetable Soup, Mess Beef and Vegetables, Concentrated Soup.
19. *Regulations and Instructions for the Medical Officers of HM Fleet*, 1825.
20. *Parliamentary Papers*, XXX, 1852, p 317.
21. *Ibid*, XLI, 1856, p 281.
22. *Ibid*, XLV, 1867-8, p 301.
23. The first tinned meat from the Admiralty Victualling Yards was issued in 1867. A girl called Fanny Adams was murdered in August of that year and her body hacked to pieces. The sailors' sense of humour suggested that this was what went into their tins and they nicknamed the tinned beef 'Sweet Fanny Adams'.
24. *Encyclopaedia Brittanica*, 7th edition, 1842, Vol 9, 'Food and Food Preservation', pp 720 ff. Sauerkraut was packed into the barrels by a man 'having on strong boots, well-washed and nicely clean'.
25. *Ibid*, p 732.
26. Hon. John Cochrane, *The Seaman's Guide*, London, 1797, p 33.
27. *Boteler's Dialogues*, p 65.
28. *Parliamentary Papers*, XXX, 1852, p 317.
29. Anon, *The Seamen of the Royal Navy, their advantages and disadvantages as viewed from the Lower Deck*, London, 1877.

CAMPING FOOD IN AMERICA: Finding Nature in Food?

Molly G. Schuchat

American Indians of the Great Plains (and other areas) used dried foods on their annual travels and to preserve their food collections. The Mormon church (founded in America in the 19th century) developed storage techniques in order to keep long-term larders against hard times and isolation.

The military have refined methods of simultaneously preserving and packaging highly nutritive and compact meals for speedy distribution.

All of these techniques have been adapted and used, some for everyday consumption and some especially for recreational camping and backpacking. The products provide an integral and economically attractive focus for the outdoor recreation market.

Between 1959 and 1978, walking, hiking, biking and other forms of trail activities consistently ranked among the ten most popular outdoor recreations in the United States.

Within those twenty years, the number and capacity of camp and picnic areas doubled (Klar & Kavanagh: 82).[1] In 1961, 71% of campers used tents and the rest some kind of recreation vehicle. But twenty years later the majority lived in recreation vehicles of some kind or another and only 48% used tents (Echelburger & McEuwen).

By 1986, of the 85% who had participated in some form of outdoor recreation at least once that year, 45% had been camping, 27% had done dayhiking and 17% had been backpackers (Market Opinion Research).

There have been concomitant changes in equipment, clothing, and in food available to and preferred by participants of all kinds. This paper examines popular practice as a function of economic and ecological circumstances. It is based on interviews with and observation of a wide spectrum of backpackers, car campers, recreation vehicle users, fishermen and short-term hikers regarding their camping food habits. For purposes of the paper, camping is a continuum from sleeping under the stars in the wilderness to spending a luxurious week in a heavily programmed summer camp for adults.

Colin Fletcher, a Hiker who has written definitive books on both the adventure and the craft of 'walking', and an Englishman by birth, muses

> ... an encounter with nature...has a powerful hold on the American imagination—an idea of independence, of self-reliance, self-sufficiency and autonomy. These are ideas that lie very close to the heart of the cultural values we prize most, and that seem to be most threatened by the style of modern, urban, industrial society.

OLD DAYS, OLD WAYS, NEW DAYS, STILL SOME OLD WAYS

The earliest settlers and wanderers in America were of course the Indians and they had immediate effects both on the first and following European settlers down to our current attitudes. Two of the most famous and still used Indian contributions to camping food are jerky and pemmican. The word jerky, is a Spanish rendering of the Peruvian Indian *charqui*. It was prepared by cutting boned and defatted venison or beef into quarter-inch slices and either dipping these in strong brine or rubbing them with salt. After the meat had been rolled up in the animal's hide for ten or twelve hours to absorb the salt and release some of the juices, it was hung to dry in the sun and then tied into easily carried bundles. One early German traveler said it looked like thick pasteboard and 'was just as easy to masticate'(Tannahill:298). Pemmican, of which more later, was the original American trail food, a ground up concoction of pounded up meat and wild berries laced with grease. It played an early and important role for the Europeans, enabling Alexander Mackenzie, a fur trader, to travel from the Atlantic coast to the Pacific in 1793. The still sold American 'candy', cracker jack, is a direct descendent of the New England Indians popcorn coated with maple syrup. The toy at the bottom of the box is the commercial marketer's addition.

Mary T. S. Schaffer traveled in the 1880s to the Canadian Rockies, in pursuit of wild flowers and animals, from the luxury of Philadelphia. She went first with her husband, and later, after his death, with friends. She wrote several volumes including *Incidents of Camp Life and Trail Life*, covering two years' exploration through the Rocky Mountains of Canada and *Old Indian Trails of the Canadian Rockies*, beginning in 1907. Schaffer explained that to prepare for her 16 weeks through the wilderness she and a female friend packed out from Philadelphia flour, baking powder, cocoa, coffee, tea, sugar, dried fruits, evaporated potatoes, beans, rice, with a week's extra rations thrown in for emergency...since they were 'going where there were no shops, only nice little opportunities for breaking and losing our few precious possessions'. (Schaffer: 18)

But she commented that there were only a few of these foods really worth having and some of them they were thankful to have tried before carting them across the continent. 'For instance, beware of the dried cabbage; no fresh air in existence will ever blow off sufficient of the odor to let it get safely to the mouth'. She also discussed 'granulose' a strongly recommended article to 'save carrying so much of that heavy and perishable, yet almost necessary, substance, sugar'. The advertisement for the granulose promised that one half ounce was equal to one ton of sugar, so they bought that amount. But the label was wrong, the dollar's worth was only equal to 30 pounds of sugar, resulting in many puddings and cakes without sweetening. They also bought dried milk and dried eggs. 'Truth compels me to state that each of the three has its limitations and to this day I wonder if that dried milk had ever seen a cow, or if any hen would acknowledge the motherhood of those dried eggs.'

Arrived in a Canadian town near the wilderness for her 1908 summer trip, 'some of us haunted outfitting shops, bought shoes warranted to turn water till worn out, invested in dried vegetables of little weight and wonderful nurturant qualities and spent hours preparing pinole', ripe corn roasted and ground fine like coffee. This last item they learned of from an old camper and prepared all day in a borrowed kitchen.

> I was tired but felt refreshed as I remembered his recommendations; 'Two tablespoons of pinole mixed in a small quantity of water will sustain life for 24 hours, and consequently is one of the most valuable foods that can be carried on the trail.' In imagination I could see our flour bags washed away in the Athabaska River, the bacon down with a drowning horse, and ourselves three hundred miles from a store, sitting around the pinole-bag, every one grateful for the thought which had prompted the addition of this valuable adjunct to our larder. (p 80)

They didn't lose any of their regular food and they packed the pinole for miles and miles until someone asked if they ever intended to use it. Thus challenged, they brought the bag out the next day.

> with a little sugar and cream it was not so bad and the bag came forth the next day...but it had a taste which hung on for hours, its consistency was that of a mouthful of sand, and its grittiness was all over you, inside and out ... on the 4th day a mere smell of it caused a howl to go up.

Next they gave it to the horses, but they also refused to eat it.

Most of the wars of man-kind have led to improvements in the packaging, preserving and distribution of food to large groups in far flung areas. Tin cans were developed early in the 19th century, refrigerator cars moved meat and other perishables perhaps fifty years later, dried foods were both old and new, but freeze dried foods are new, while many compressed nutrition packages, like the incredibly rich US Army K and C Rations were only developed in the 1940s. No soldier and few civilians from the Second World War can ever forget canned Spam—still the most widely sold canned meat in the USA. Commercially frozen foods and home freezers developed like Siamese twins while the space age has brought many innovations. The most widely known is Tang, the essence of orange juice to which you just add water and stir. One of the most delightful is dripless ice-cream (blue wafers that melt in the mouth) concocted by a nutritionist for the Apollo flight program.

Today over four million children and youth participate in summer camp yearly, joined by another four million children and adults for day and resident camps of many kinds—privately owned, social agency or religiously affiliated, institutional, and special purpose camps, covering more than a million acres of prime recreational land across the U.S. They generally eat the same institutional food available in school and senior lunch programs during the winter—with the addition of weenie roasts and other evening campfire activities.

According to the American Camping Association(iv) organized camping began in 1861 when Frederick William Gunn and his students trekked to Welch's Point in Milford, Connecticut to camp on the beach, from their school in Washington, Connecticut. The Association itself was founded in 1910, the same year that the Boy Scouts of America began activities in the US. The fact that 31 million Boy Scout Manuals have sold since then, attest to the vibrancy of that movement in training generations of American youth in outdoor ways. One of the early tasks to be accomplished in the succession of badges to attain First Class Scout Status is the ability to prepare a cooked meal in the woods but new times have greatly modified that meal.[2]

It is an American tradition/myth that young men set out, packs upon their backs, and 'go West' to make their fortune. According to Adler it was a European tradition from the Middle Ages on that young men went on pilgrimages not for religious purposes but to look for work, and young artisans would travel from town to town, sojourning with local craftsmen and honing their skills until they found a permanent base or returned home. Adler suggests that this served as a ritual aid in accomplishing the separation from home and family required by western styles of adulthood (until well into the 19th century when railroads changed travel patterns) while offering young men the opportunity for sightseeing, adventure and education; and she reminds us that this pattern of labor-related travel of lower classes developed as early and as fully as the recreational and cultural travel of privileged youth.

In the (American) world since 1960, certainly, the road culture of youth provides an example of an upwardly mobile rather than a downwardly mobile cultural tradition, suggests Adler, 'with middle class youth increasingly adopting a mock version of this earlier working class travel pattern'.

MARKETING AND MODERNITY
The food buyer for a regional chain of trail outfitters, in the business for the last 6 years but 'backpacking all my life', observes that the pattern has changed enormously in the last years, for a variety of reasons - he certainly does not have the time for extended trips any longer, and usually focuses on two or three days at a time. His job sometimes permits him mid-week time off, which is much better for such journeys, because the woods are full of weekend hikers and campers, which destroys so many of the reasons for going out there in the first place. And recently he has spent more time using trail bikes than his feet. He still carries his pack on his back, 'but believe me, seated on the bike you just don't feel it like you do hiking'.

'Camping', he observes, 'used to be a cheap sport, but the wonderful equipment now available is anything but inexpensive. Of course, if you figure the cost over a long period of time, it is still quite reasonable, but it is not general purpose equipment either, each type of trip demands its own.' Nowadays, he says, the baby boomers who did all the backpacking in their youth are established professionals and have young families. They are more likely to rent a cabin and do day hikes. This of course has a profound effect on the type of food they buy.

The regional chain for which he works began 13 years ago as one store in a rather distant suburb of Washington, nearer the Appalachian Trail than the city itself. It has just opened its 13th outlet. A venerable Washington State cooperative outfitter, REI, simultaneously opened its first Washington, DC outlet. REI, the East Coast L.L. Bean company in Maine and other once local stores are now national catalog stores as well as retailing chains. One accompaniment of this growth in retailing has been an ever growing focus on outdoor clothing and some decline in the other outdoor merchandise offered. At the same time there has been quite a decrease in the importance of camping food as items in stores and even in catalogs! Supermarkets can supply, today, almost all of the products needed and very few people make anything even resembling the pinole described so painstakingly by Mrs Schaffer.

In the mid 1980s a good measure of the change in camping needs and attitudes is offered by the Sierra Cup, named for a wilderness and environmental guardian society whose members were originally rugged backpackers of the Western Mountains. The 10 ounce Sierra cup, very lightweight stainless steel, has a wire handle that fastens under the lip of the cup and completely around it for strength. The wire handle fits over one's belt and the cup lies snug against one's jeans, instantly available for dipping into the nearest mountain stream for that wonderful fresh cool water. Now the Sierra Cup, ever more a symbol of one's outdoors connectedness, is also available in considerably heavier brass and silver and even gold-plated versions (engraved monogram or message extra) as well as even lighter plastic. The plastic is cheaper, tough and cleanable, measure-marked, but with less stability and, of course, one cannot cook in the plastic cups.

And today no one dips the cup into the cool mountain stream, not in the high Sierras and Rocky Mountains or the lower but still remote Adirondacks on the East Coast. The scourge of giardia has spread to all these streams. Sometimes called 'The Beavers' Revenge', this protozoan, also known as backpackers' disease, comes complete with diarrhea, cramps, visible bloating, weight loss, nasty burps and loss of appetite and vomiting.

And most places no-one lights camp fires. Colin Fletcher, that dean of Walkers, says that 'half the fun of camping, for so many people, is a fire—for cheer, warmth and cooking and drying out self and clothes'. For groups, campfires multiply the pleasures of the well earned rest and the opportunity to exchange tales of past triumphs and adventures. But as the woods have become filled with people the loose wood is gone and the risk of inciting a destructive fire so great that the National Parks have banned them, and many of the other wilderness areas as well. Fortunately, at about the time that fires began to come under bans, new stoves became available for backpackers and others, less bulky and even more reliable than the earlier stoves. There is no time consuming effort of hunting wood, it only takes a second to light the fire, pot exteriors stay clean, and in rain or snow one can cook and eat in the cocoon of a sleeping bag. There is a weight problem in having to carry alcohol, gas or butane cartridges.

Fletcher, ever the enthusiast and optimist, says that cooking over the open fire may be time honored, but it cuts him off from the night. What he really wants from the wilderness is to reestablish a sense of unity with the rest of the world, rocks and trees and animals and sky and its stars. 'The Campfire, although it charms, disrupts my sense of inclusion.'

AS TO THE FOOD
All of the equipment needed to produce home dried, home pulverized, home mixed and home cooked food to take along on the trail is still available as can be seen from the various company mail catalogs. Most campers do not indulge in all of the work required to grow their own, dry their own and package their own and most campers do not need to, because they are not going on long treks like Fletcher's walk through the Grand Canyon. And a few hikers take no stoves, expect no fires, and eat only what needs no trail preparation at all.

Two weeks backpacking in Alaska viewing the wildlife in an out of the way valley included, wrote a reporter, 'two pounds of food per day per person, which meant each of us would carry 24 pounds above and beyond our own personal clothing, sleeping bags and common gear. The food was mostly dry or freeze-dried, plus crackers, cheese, jam, honey, almond paste and 'gorp'—a high-energy trail mix of chocolate, nuts and raisins. The total cost of food and fuel for four people for two weeks: $380'. (Bohlen)

Very reasonable? Yes, but getting there once arrived in Anchorage, Alaska was $111 to Fairbanks and back, a charter flight from Katovik to Kongakut River at $270 per person round trip. If one doesn't go with experienced Alaska outback travelers, a qualified outfitter is a must. This runs to about $700 a person, but includes the food and cooking equipment, plus the transportation. Of course this is one of those incredible peak experiences ever less achievable even in the distant outbacks of the world.

In a Marketing Survey prepared for the President's Commission on Americans Outdoors (1987) motivations for outdoor recreation were found to cluster in several groups. Across the spectrum of persons interested in camping are those interested in observing nature (31% of the adults). Then there is a collection of persons devoted to fishing, hunting and horsepower, who use RVs motor boating and off-road vehicles, all of which also involve 'camping', but in quite a different way. The water enthusiasts who sail and canoe are again in a different category. Factor analysis of the survey motivations indicated that competitive persons, who do the fishing, hunting and horsepower activities are predominantly male; the nature lovers, who are not competitive, are half male and half female. The survey only included persons over 18, which does not count the boy (and girl) scouts whose paths cross these adults on the Snake River, the Appalachian trail and on cross country biking trips—a group who account for a tremendous amount of the camping in America. All of the types interact with the motivations to be social, to have fun with friends, and to be with family together (Market Opinion Research, New York, 1986 passim).

One of the people whom I interviewed, who has kayaked and canoed inside the Arctic Circle and kayaked and hiked and camped in most of the North American Wilderness areas spoke about the need to equip and pack and think differently depending on what kind of experience, as well as the duration, planned. She has three basic categories, car camping, back packing and white water activities. Which she does depends on time and money, and she is, unfortunately, usually short of one or the other. Given the money and lack of time, it is car camping, or perhaps parking the car and going off for two days from where she has driven. Car camping means you can take just about anything you want to eat and you don't have to worry about carrying any of it. Ice cubes and frozen foods also keep that way in a portable cooler. Organized campgrounds, both public and private, usually have fireplaces and/or gas grilles, hot showers, and hookups (particularly in commercial campgrounds) to attach equipment in Recreation vehicles. Commercial campgrounds also accept pets, which National and Some State Parks do not. So it is easy to stock up on food at both the trail stores and the grocery store on Friday afternoon on your way out to a camping weekend. Water travel has a different set of priorities.

At the Marina and camping store in the Everglades National Park in Florida all manner of fishing and camping equipment is available for the mangrove swamps and Florida Bay. Although dehydrated and other prepared camping foods are available, the shelves emphasize canned goods—stews, hashes, even tuna fish and chicken salad in cans. Motor and sail boats both have plenty of room for storage and weight is not a problem. What is, parked in a salty bay, is fresh water for drinking and cooking—but again, boats usually have lots of room for all one needs to carry. Canoe trips, of course, do not offer as much storage space, but still enough. For campsites in the Everglades, one reserves ahead via Tiketron system (which is also a way of getting city cultural tickets nationwide) paying for the space (and the Tiketron handling fee) by credit card. In the Everglades, where the bordering land is very fragile, camping is not allowed on shore, but the Park Service has constructed platforms over the water, complete with enclosed portable toilets. Cooking facilities are as in more landlocked wilderness areas.

Some campers (but few comparatively speaking) still prepare ahead and, if not going by horse, mule or the newly fashionable llama, keeping the backpack manageable is still foremost. But most trips are weekends or four days here and there, and supplies are now available not very far from the trail, in the little villages that adjoin even backcountry. Even on my two week wilderness ride 15 years ago (the wonderful meals, based on dehydrated foods and fresh ones—including the cut-throat trout we caught in mountain streams and lakes, with everything else packed in by our accompanying mules) it appeared that we were deep in the wilderness, with nothing around but the usual small animals, many deer and an occasional moose. However, two days from the end of our trek we ran out of bread.

No make-do would satisfy our outfitter so one of the cooks rode into the nearest town to bring back bread for our next day's lunch—four hours each way. Meanwhile, we riders moved on, all unknowing, deep in the wilds, light years away from civilization.

But now, not just white bread, but the country's most luxurious frozen croissants as well as local home made breads are available at that distance, or even less. And for cabin vacations—in Parks or private facilities— everything is available at the general store down the highway—frozen Norwegian or Pacific salmon in the byways of the Massanutan mountains of 'Wild, Wonderful, West Virginia' (the state tourist motto stamped on brochures and blasted on roadside signs—but not in the wilderness areas).

This same availability has also changed the nature of what constitutes a cheap family vacation. On the interstates, sometimes in the most beautiful country, where no roadside signs are permitted, a forest of gas, lodging and McDonald golden arches signs beckon one for a night's rest from the rise of the hill near the next exit. Two double beds and a fold out sofa are standard equipment, so a family can sleep more reasonably for an occasional trip than buying the good equipment 'necessary' for satisfactory 'cheap' family camping. And who wants to cook?

Actually, many people do—they rent cabins and acquire the necessary food and prepare gourmet dinners to end the day following a day-hike. For that they carried peanut butter sandwiches and 'gorp' in their day pack, and a canteen. The 'gorp' has long been a favorite home made and/or purchased trail food. It is usually based on dried fruit and nuts, sometimes with chocolate or carob. One recipe is:raisins, dates, coconut, figs, prunes, pecans, walnuts and filberts ground together in the food chopper. Pack the result into a one inch metal tube, lay on waxed paper and push the cylindrical rod of 'gorp' out. Wrap it in the paper, then in foil and stow in the refrigerator. On the hike, peel back the foil and eat as one would a banana.

Many of these cabin campers backpacked, first, during the heady 1960s when, free spirits all, they lived on grains in the wilderness. Even now, although no longer vegetarians, they eat as healthfully as possible with fish and occasional meat and few, if any sweets, except when out hiking and camping. Then they indulge in chocolate bars both for energy and for a reward for their heavy physical activity. Now from their cabins on their vacations from two careers and with their little children, they do not have a shortage of money, but time. Still the end in view is the great outdoors.

One, who has camped extensively both in Europe and in the US, says there is a major continental difference. In Europe transportation by bus is available from place to place and camping is a reasonable way of seeing the sights. In America the camping in the great outdoors is the goal, not the means to more rounded sightseeing. Otherwise people are more often taking their recreation vehicles and parking outside of town or in their friends' driveways. And in America most camping sites are reached more or less by auto—or van—or at the least by flying to a reasonably nearby city, renting a car and then parking it and taking on their packs, or unloading their bikes, or—most recently—placing their packs on their trail bikes. And on their backs or on their bikes or llama they pack not beans but foods for simmering, steaming, sauteing, baking and other gourmet preparations.

Campgrounds are big business—the largest, Kampgrounds of America and Leisure Systems, both franchisers, include mostly transients' facilities.Smaller campgrounds, which cannot compete with these nationwide systems' advertising, have switched to patterning themselves after condominiums and cooperatives, selling individual lots or a share in the property. Others operate as private clubs whose members then use the extensive and elaborate facilities—but camp, mostly in their Recreation Vehicles, although a few still set up tents adjoining their cars and trucks. Restaurants are on the grounds and there is very little camping food eaten and less hiking.

Meanwhile, back at the ranch! In America's home backyards and on apartment balconies the gas and charcoal (bought at the supermarket in large sacks) grilles are ubiquitous and it is there that the traditional steaks, hot dogs and hamburgers continue their outdoor lives. Americans blessed with the most elaborate kitchens in the world as part of the expected way of life, prepare their traditional camping food for home consumption. Barbecuing or grilling meat over an open fire may be a primitive technique, still it is a popular seasonal way of life in America.

And yet not always. Sometimes the lure of the 'real' outdoors calls. Those who can afford it can then go trout fishing in the mountains for a 'different, back to nature' experience. A few may spend $2,500 on rods, reels and other equipment and another $800 on a three-day course on fly-fishing. Or they may buy recreational products from kites and power boats to camping gear and croquet sets, put it all in the car and go off somewhere, anywhere with the very latest in everything for the wilderness, including prepared gourmet camping food that just needs the pouch heated in water. Or carrying their relentlessly vegetarian bear-valley pemmican. Or even eating President Reagan's favorite 'snack' food, bee pollen bars, filled with peanuts and raisins instead of home-made gorp.

CONCLUSION

The camping experience is one of finding nature, for a day or a weekend or a few weeks. For the shorter hauls it is the appearance of wilderness that is important, the breathing of fresh air, the light and shadows in the woods, sighting forest animals, finding a clump of wild food—these last becoming less and less likely, even in the deep wilderness. For those who want fresh green food, the simplest thing these days is to take along your own beans and sprout them on top of your backpack. Otherwise the grazing is hard to come by, or else you share it with the bears in the back country, which I suppose is part of the experience of nature and danger, and the feeling of 'traditional' camp food eaten when and how we were meant to by nature, at the end of the physically taxing search for a food source. As one informant put it, 'food that is good for motion, present tense, the result of foraging'. But he also suggested that today's foraging is the prepared 20th century high-powered food available from the stores: dehydrated beef stroganoff rather than beef jerky.

I think that for some adults food in the outdoors is an attempt to recreate childhood, when popcorn, marshmallows and peanuts roasted over the campfire rekindle the feeling of the innocence of days long gone, ontogeny rather than phylogeny.

And for still others in this post-industrial high-stress urban world, it is the preparation, dressing for the occasion, the careful organization of props, not the immediacy of the situation, that recalls the dream. All the world's a stage and all the people merely players, acting as intensely in their primitive roles as in their professional ones, with all the tricks of the trade necessary and no improvisation understood.

Some see 'civilization shot' by a return to the great outdoors, but others perceive that is the very 10,000 years of civilization that has permitted us to play in the wilderness.

REFERENCES

Adler, Judith, 'Youth on the Road: Reflections on the history of tramping'*Annals of Tourism Research* 12#3, (1985), pp335-370.

American Camping Association, Parents' Guide to Accredited Camps, Martinsville Indiana, 1987.

Bohlen, Janet Trowbridge, 'Backpacking Above the Arctic Circle',*The Washington Post Travel Section*, 3 May 1987, E1 & ff.

Clawson, Marion and Carlton S. Van Doren, *Statistics on Outdoor Recreation*, Washington, DC, Resources for the Future, 1984.

Cook, Marc, *The Wilderness Cure*, New York, William Wood & Co., 1881.

Echelberger, Herbert E. & Douglas N. McEuwen, 'Activities' in*Literature Review, Appendix to The President's Commission on Americans Outdoors* U.S. Government Printing Office, Washington, DC, 1987, p 71 & ff.

Fletcher, Colin, *The Complete Walker III: The Joys and techniques of hiking and backpacking*, New York, Alfred A. Knopf, 1984.

Green, Arnold W, *Recreation, Leisure, and Politics*, New York, McGraw Hill, 1964.

Hillcourt, William, *The Official Boy Scout Handbook*, Irving, Texas, Boy Scouts of America, 9th Edition, October, 1985.

Kaplan, Max, 'Leisure in America' in Pauline Madow (ed), *Recreation in America*, The Reference Shelf vol. 37#2, New York City, H.W. Wilson Co, 1965.

Klar, Jr., L.R. & Jean S. Kavanagh, in *Literature Review, Appendix to the President's Commission on Americans Outdoors* Washington, DC, U.S. Government Printing Office, p 82 & ff.

Goodman, Paul, 'The Mass Leisure Class',*Esquire*, July, 1959.

Market Opinion Research: *Participation in Outdoor Recreation Among American Adults and the Motivations which Drive Participation*, New York City, underwritten by the National Geographic Association, 1986.

Rinehart, Mary Roberts, *Through Glacier Park in 1915*, Boulder, Colorado, Roberts Rinehart, Inc Publishers, (originally published 1916 ,P.F. Collier & Son, Inc).

Root, Waverley & Richard de Rochemont, *Eating in America: a history*, New York, William Morrow and Company, Inc, 1976.

Sax, Joseph L., 'Albright Lecture' quoted in Colin Fletcher, *The Complete Walker* (1984), 46.

Schaffer, Mary T.S., *A Hunter of Peace and Old Indian Trails of the Canadian Rockies*, Banff, Alberta, Canada, The Whyte Foundation, 1980 (*Old Indian Trails* originally published by G.P. Putnam's Sons, NY, 1911)

Tannahill, Reay, *Food in History*, New York, Stein & Day, 1973.

Unruh Jr, John D., *The Plains Across: Overland Emigrants and the Trans-Mississippi West*, 1840-60, Urbana, University of Illinois Press, 1979.

NOTES

Attendance at this Conference was made possible by a travel grant administered by the American Council of Learned Societies and funded by the John D. and Catherine T. MacArthur Foundation and the National Endowment for the Humanities. I am grateful to them all for the support.

1. Camp and picnic areas increased from 286,131 person capacity in 1960 to 566,477 person capacity in 1980. But the number of areas for camp and picnics increased from 5,067 to 6,326, only a 25% increase in facilities.

2. Cooking skill award from 1985 edition of Boy Scout Manual pp 542-3.
   COOKING SKILL AWARD
   1. Show you know how to buy food by doing the following
      a. plan a balanced menu for 3 meals: breakfast, lunch, and supper
      b. make a food list based on your plan for a patrol of 8 scouts
      c. visit a grocery store and price your food list
      d. figure out what the cost for each Scout would be
   2. Sharpen a knife and an ax properly and give rules for their safe use.
   3. Use a knife, ax, and saw correctly to prepare tinder, kindling, and firewood.
   4. a. locate and prepare a suitable fire site,
      b. build and light a cooking fire using not more than two matches.
   5. a. in the outdoors, cook, without utensils, a simple meal. Use raw meat (or fish or poultry) and at least one raw vegetable, and bread (twist or ashbread),
      b. in the outdoors, prepare, from raw, dried, or dehydrated food, for yourself and two others: (1) a complete breakfast of fruit, hot cooked cereal, hot beverage, and meat and eggs (or pancakes), and (2) a complete dinner or supper of meat (or fish, or poultry) at least one vegetable, dessert, and bread (biscuit or bannock).
   6. After each cooking, properly dispose of garbage, clean utensils, and leave a clean cooking area.
   NOTE: When laws do not let you do some of these tests, they may be changed by your Scoutmaster to meet the laws.

RAGS TO RICHES: Salmon's Rise from Poor Man's Food to Gourmand's Delicacy[1]

Robert J. Theodoratus
with the assistance of Alice J. Parker

In aboriginal times 'smoked salmon' was the staple food for the Skagit and other Northwest Coast Indian peoples. After 1800, Europeans and Americans began to commercially exploit the vast fishery resources in this region. They introduced brining and salt packing. This was followed by canning in tins, and in time, quick freezing. For a variety of historical and cultural reasons, the early settlers and their immediate descendants rejected the use of smoked fish although they would eat canned salmon. In the 1940s and 1950s an effective rural educational extension program reintroduced fish smoking into the local scene. A rising respect for local Indians also aided this process. More recently, low cost home electric fish smokers have been the primary agent in popularizing home smoking as a means of preserving salmon and other fish both in rural and urban areas. However, in this historical process of continuity and change, what smoked salmon once was and what it is today are two very different foods.

When I was a graduate student in the 1950s, my parents, who lived in the Upper Skagit River Valley about 90 miles north of Seattle, Washington, would often send me large parcels of home-smoked salmon. To me, a small amount of salmon in this form was nice as a snack and no more. This created a minor moral dilemma; what to do with most of it since I had grown up in a culture which emphasized that one should not waste food. Fortunately, two of my friends were more than willing to help me soothe my conscience by taking most of it: one a Jewish-American and the other Polish. Today there would have been much greater competition for my 'care' packages from home. A change in attitude which led me to examine how changing values and technologies in preservation have changed the way people in the Skagit Valley value salmon as a food.

Often such a topic first appears to be a simple sequence of historical facts but turns into a complex set of interacting sociocultural events and processes. This is what happened as I began to examine some of the historical, social, ecological and technological changes which have been responsible in the transformation of salmon from a poor man's, and in some forms even a despised food, into a gourmet delicacy in the United States in general and the Skagit River Valley of the Northwest Coast of North America in particular.

Salmon was a cornerstone of the aboriginal Indian diets. Before Europeans arrived, Northwest Coast Indians seasonally caught, processed, and preserved vast stores of salmon[2] for use throughout the year. While the different species of salmon were migrating upstream, they were barbecued over open fires, cooked in earth ovens or stone boiled. However, most of the vast catch was dried and/or smoked for use during the rest of the year.

The usual preservation process included butchering the salmon, then filleting and/or cutting the fish into strips before hanging the pieces on drying racks which were exposed to the air, wind, and some sunlight through conifer bough roofs. In cool weather as autumn progressed, smouldering fires, usually of alder wood, were built under the drying platforms to counter the humidity caused by the increasingly common light rains. The salmon were also cut into narrower strips to facilitate the drying process. The smouldering alder wood fires added a preferred flavor as well as repelling insects and scavenging birds. No salt was used in this process.

This relatively simple process produced two forms of preserved or 'smoked' salmon: 'light' or 'soft smoked' and 'hard smoked'. The first was only partly dried and lightly smoked with some moisture remaining in the flesh. The strips were also placed lower and directly over the fires in order to partially cook the flesh. This had to be consumed in a few weeks or one risked much spoilage (Elmedorf 1960: 121; People of 'Ksan 1980: 22; Smith 1940: 238-9).

For 'hard or dry smoked' salmon, lower temperatures, a longer time and more exposure to smoke was necessary. This form was the staple for long-term storage into the early summer of the coming year. The strips were cut thinner to facilitate drying, ease of storage and to retard molding. In the smoking and drying process each piece was frequently observed and turned on the racks and knife-cuts were made into the pieces to speed up drying and the penetrating action of the smoke. Women would also periodically rub and squeeze each piece in order to break the fibers and facilitate drying and smoke penetration. It usually took from one week to ten days to hard smoke fish. After the pieces were dry, the strips were tied into bundles and stored up in the rafters over the open hearth cooking areas where it was warm, dry and somewhat smokey (Smith 1940: 238-9; Singh 1966: 50-1; Underhill 1945: 22-4).

Special care had to be taken in obtaining salmon for hard smoking or drying. The salmon were caught some distance upstream; in the Skagit Valley, this was usually 40-50 miles upstream near the spawning grounds. Since Pacific salmon cease to feed once they enter fresh water, no body fat remains by the time they spawn. All die after spawning. The flesh of these fish keeps best since there is no fat to turn rancid. I should also add that the Indians also smoked salmon eggs, heads, and fins along with halibut, clams, oysters, and mussels.

By 1800, American, British and Russian trading companies had begun exploiting the salmon for markets in Hawaii, Europe, the Far East as well as the Eastern United States. They introduced the northern European technique of brining by packing the salmon fillets between layers of salt in wooden barrels. This preservation process draws the moisture out of the flesh to produce its own brine. Much of the fish shipped to Europe and eastern American cities was usually soaked overnight to remove the excess salt. It was then lightly smoked for 10-12 hours to produce the type of salmon then in demand in restaurants.

This type of salty salmon evolved into the increasingly popular *lox* served with bagels and creamed cheese in the massive New York Jewish community in the early part of this century, helped by the fact that the salt made the salmon *Kosher* by drawing out the blood. In time *lox* became the commercial type of smoked salmon sold in urban delicatessens and markets all over the United States. For the average American today this *is* what is meant by smoked salmon (Mariani 1983: 235; Rosten 1968; Schuchat, personal communication).

However popular *lox* was to become, vacuum packing in tinned cans slowly began to dominate the commercial production and distribution of salmon by the end of the 19th century. This continued into the post World War II era. The then natural abundance of salmon, ease and efficiency of catching salmon, and the low cost of processing and shipping salmon led to it becoming an abundant, low-cost food for Americans. In the Pacific Northwest, it became easier and cheaper to purchase tinned fish than to fish for it even when abundant and nearby. At the same time, federal and state laws were enacted banning salmon fishing in upstream waters near spawning grounds. As time went on these restrictions were more stringently enforced, thus creating much conflict between contemporary Indians and game wardens over whether or not they could catch the fatless upstream salmon for hard smoked fish.

Since World War I, an increasing proportion of the annual salmon catch has been preserved by commercial quick freezing. Most of this frozen salmon has been marketed outside of the Pacific Northwest. What was and is marketed locally is primarily sold among urban-suburban populations. In smaller communities in the Pacific Northwest, the demand for frozen salmon has been minimal.

In the Skagit Valley (c 1880-1920), the early settlers attitudes towards salmon were influenced by their attitudes towards the native American Indian population. This early migration came mostly from interior farming communities, whether from the Midwest and Middle Atlantic states and the Piedmont of North Carolina or Inland Sweden, Norway, Germany, and other European countries. Instead of the dashing Plains Indians with their horses and feather head-dresses, they found *siwashes*, a local corruption from the French term *sauvage* (savage or primitive) which was applied to the coastal Indians by early French-Canadian fur trappers.

The Skagit Indians lived in plank houses and ate fish. Worse, they seasoned most of their foods with fish oil. To the new settlers, the natives reeked of fish, and *siwash* became a truly derogatory label. Whites carefully avoided eating anything or doing anything Indians did until around the 1950s when attitudes began to change. This attitude even penetrated the curriculum of the schools where they included information on the Indians from the Plains, Eastern Woodlands and the Pueblo Indians but never the coastal Indians who were native to the area.

Olfactory and visual experiences reinforced this negative attitude towards salmon. In the early decades and into the 1940s, the sight and smell of millions of dying and rotting fish filled the banks and shores of streams in the valley each autumn. Few people wanted to even think about eating fish during those months let alone smoke them. Those 'whites' who operated smoke houses, only smoked pork and pork products.

Immigrants from other countries soon began to share the local American attitudes towards eating salmon. Most of them came from interior or inland regions where river fish were an occasional food supplement rather than a staple. Salmon were viewed as an annual plague.

This basic pattern of local salmon use/non-use continued into the 1940s. The Upper Skagit Indian families who remained continued to catch and smoke salmon for their own use. However, their staples became potatoes, bacon, coffee, pancakes, etc. Salmon declined in importance. The Indians were increasingly harrassed by local fish and game officials for spearing salmon near the headwaters of local streams. What little hard-smoked salmon they made became an illegal commodity. Whites scorned those upstream salmon because they were old, tough, tasteless and almost ready to die. They were useless for baking or frying.

For local whites most of the salmon consumed was tinned and from local stores. Its cost remained minimal, it was available the year around, could be prepared quickly and was mild flavored. It was normally served and eaten as one would do for meat. Fresh salmon was not commonly found in local small town markets: if one desired or needed to eat fish (e.g. Pre-Vatican II Catholics) most preferred halibut, smelt, oysters or canned salmon. Before 1940 few whites were even aware of the existence of hard smoked salmon. The salty lightly smoked salmon (*lox*) was only to be found in delicatessens and a few Jewish kosher food shops in the large cities in the Pacific Northwest. Tinned salmon reigned supreme in the valley.

In order to understand how and why people in the valley eventually began to consume and even smoke salmon themselves, we must look at other developments outside of the valley.

In recent generations, state and federal extension/educational services have been an important force in improving economic, nutritional and educational conditions in rural and small town America. Much of this has occurred through published circulars, leaflets, booklets, etc[3] on agricultural improvement, nutrition, practical mechanization, etc. In the past such items were available at a minimal cost or free from the issuing governmental agency, political office holders, or one's county agricultural agent.

The greatest surge of this information flow occurred during the Great Depression years (1930s) as a part of rural reconstruction programs. In those years agencies and individuals helped local rural families construct home freezing units, smoke houses, and better food storage units. There were earlier efforts beginning in 1917. However, their total orientation was towards New England, the Middle Atlantic States and riverine fish in the

Mississippi Valley. The Pacific Northwest was barely noted (United States Department of Commerce ... 1917).

Other leaflets and books were published throughout the 1940s. Though these were still primarily for eastern North America, they did include increasing amounts of information which were either on the Pacific Northwest or could be adapted to fish of this region. Finally, in 1947, 1969 and 1978, the State of Washington Bureau of Fisheries issued its own booklets on smokehouses and smoke curing of North Pacific species of fish. All instructions included the European technique of soaking the fish in brine overnight before smoking. Though focussed primarily on salmon, these also provided information on trout, smelt, clams and oysters. Recipes and recommendations for the use of home smoked fish as an appetizer in salads, sandwiches or as a main dish were included. All three were oriented towards the urbanite, rural dweller and the sports fisherman. It was through these that an interest was aroused in home-smoking in this region. It also helped many overcome some traditional food prejudices (Anderson and Pederson 1947; Berg 1969, 1978).

The directions and plans for the construction of simple and practical smokehouses at a low cost were readily acceptable on the local level. Most of the designs and techniques were those already in use in New England and the Maritime Provinces of Canada. One was simply to convert an old ice box or refrigerator as follows.

> Drill some holes at the top to allow the smoke to escape. Use a hotplate at the bottom, preferably one with a temperature control. This provides a better control for hot smoking. Run a cord through the drain if an icebox is used. Into an old large pot place a few pieces of alder to smoulder (Berg 1969: 3).

Many were built then. However, within a few years the state health officials had to issue health warnings about the metal rods in old refrigerators as they contained cadmium. Similar cautions were issued for the use of galvanized wire racks because of the chemical reaction of the zinc to salt and creosote.

Another type of smoker recommended could be made from a steel 50 gallon oil or gasoline drum with an inverted metal washtub on top to retain the smoke. Others were of wooden barrels or cheap lumber. For all of these the emphasis was placed on minimal cost, efficiency, labor-saving, low fire hazzard, no waste or spoilage and adaptability to most species and sizes of fish (Berg 1969: 3).

Additional information was provided on properly hanging the fish in the smoker (including sketches), brine preparation, smoking durations related to differing weather conditions, storage, pit smokers for fishermen, kippering and cooking. With these available, smoking fish became increasingly common (Anderson and Pederson 1947; Berg 1969, 1978).

Another important change in the technology of salmon smoking was the increasing availability of locally privately rentable cold storage units and home freezers after 1946. These enabled local fishermen to immediately freeze salmon or other fish until needed for cooking or smoking. Today the general practice is to quick-freeze one's fish until enough has been accumulated or the person has the time or inclination to smoke it at his/her own convenience. This can be any season of the year. Neighbors can even pool their fish for a common or shared smoking.

Smoking salmon at one's own convenience has become of critical importance because of the drastically reduced runs and catches on the Skagit and other rivers. The reduced runs are a combined result of many accumulating factors: over-fishing for many decades in the rivers and ocean, construction of high dams for hydro-electric power, water pollution, destructive logging practices on mountain sides consisting of glacial till. This has caused severe erosion with the resulting gravel blocking creek mouths during salmon spawning times. The salmon cannot return to where they were hatched and thus continue to die without spawning. The reduced salmon runs along with shortened fishing seasons, and other restrictions has led to increasing prices for salmon. What was once an inexpensive food has now become scarce and costly. For many, salmon are now near luxuries.

Another aspect has been the so-called 'Boldt Decision' in a United States federal court. In this legal decision all local American Indians in the Pacific Northwest were given a larger share of the catch and many netting restrictions removed on rivers. Limits for white fishermen were severely reduced. With fewer fish to be caught, freezing for later smoking has again become increasingly common by necessity.

Since World War II the status and respect for the local Skagit Indians has greatly improved. In the early 1950s they first began to participate in one local town's annual celebration by operating an 'Indian Salmon Barbecue[1]. From its inception this rapidly became one of the most popular places to eat. They also sold soft smoked salmon to local people and to tourists. However, they have no hard smoked salmon for sale. This is but one aspect of a regional 'rediscovery' and revival of Northwest Coast Indian arts, crafts, ceremonies and foods.

Around 1960 an influx of older urban outsiders, building and purchasing vacation or retirement homes, began to occur in this valley. These people brought many of their urban food preferences with them including the heavily brined light smoked salmon (*lox*) traditional by now in urban shops and markets. Since 1960, one local supermarket has continually stocked this type of 'smoked' salmon for parties. However, almost all is sold to those of outside origin.

The factory built electrical home smoker, which first appeared around 1965, has probably done more to popularize smoked foods both locally and regionally. These mass-produced smokers have appealed to 'gourmet cooks', 'sportsmen', house-holders and all to whom gadgets in general have an appeal. They have to a large extent replaced the old large smokehouses and have popularized home smoking among ordinary people. This has been because of their low cost, small size, portability, simplicity of design and operation, efficiency and the fact that they can be easily stored in one's garage.

The advertising or promotional literature for these electric smokers have facilitated a rapid growth in their use, but often creates ironic images. The model called 'Little Chief', made by Luhr Jensen and Company at Hood River, Oregon (along the Columbia River) does not depict a Northwest Coast Indian. The 'little chief' is a seated Plains Indian wearing a feather head-dress in front of a Plains Tipi and has a plump juvenile face. Stereotypes definitely have a long life in American advertising.

The writers of the enclosed brochure which instructs new purchasers do try to give a brief promotional history of smoking food as a form of preservation. They inform the purchaser that smoke curing of fish and meat goes back to the neolithic and was done by the Chinese as early as 1000 BC. They also proudly claim that this smoker is (was) fully endorsed by James Beard, the late, renowned cookbook author and gourmet who *was* born in Portland, Oregon. Throughout the booklet the term 'old fashioned' is frequently used as are such folksy-fake southern US dialect terms as 'smokin' and 'doneness'. A number of recipes are included to help the novice produce edible results the first time they try the smoker.

Except for the seasonings and the types of wood chips used, the process for using the smoker is similar in most recipes. The heating element at the bottom of the smoker is basically a modified electric hot-plate which is set for 165°F (74°C). A pan of wood chips (hickory, alder, apple or cherry) is placed on the heating element and the heating element ignites them. Meat or fish or other foods which are seasoned to taste are put on the metal racks and cooked in the smoke.

Most people who smoke fish today freeze their catch until it is convenient to smoke. The smoker is taken into the back yard and an electric extension cord run out to it. The usual manner of preparation includes soaking the thawed fish in brine overnight, rinsing the excess salt off, and slightly drying it before it is placed on the racks. The fish is then smoked for between 14-16 hours or until the individual thinks it is done. Some of the smoked fish is refrigerated for use.

Surpluses which cannot be eaten in a month are usually canned in pint-sized wide-mouthed glass sealers which are processed in a pressure cooker. In this way the smoked fish will safely last up to four years. This is the preferred way to preserve smoked fish; preferred to freezing because smoked fish is easily freezer-burned and quality-wise does not last more than three to four months. In this way modern fans of smoked fish can be sure to have a reserve stock of 'smoked fish' in case the next year's run and catch of salmon is poor.

BIBLIOGRAPHY

Anderson, Clarence L. and Robert K. Pederson, 1947 *The Smoke-Curing of Fish and the Application of a Controlled Method to the Process*, Technological Report no 1, State of Washington, Dept. of Fisheries, Olympia, WA.

Berg, Iola I., 1969 *Smokehouses and the Smoke Curing of Fish*, State of Washington, Dept. of Fisheries, Olympia, WA.

1978 *Smokehouses and the Smoke Curing of Fish*, new edition, State of Washington, Dept. of Fisheries, Olympia, WA.

Cohen, Fay G., 1986 *Treaties on Trial: The Continuing Controversy over Northwest Indian Fishing Rights* University of Washington Press, Seattle.

Dwelley, Charles M. and Helen May Dwelley (ed), 1979 *Skagit Memories. Stories of the Settlement Years as Written by the Pioneers Themselves* Skagit County Historical Society, Mount Vernon, WA.

Elmendorf, William W., 1960 *The Structure of Twana Society*, Research Studies, Monographic Suppl. no 2, Washington State University, Pullman, WA.

Jarvis, Norman D., 1943 (June) *Home Preservation of Fishery Products: Salting, Smoking, and Other Methods of Curing Fish at Home*, U S Dept. of the Interior, Fish and Wildlife Service, Fishery Leaflet 18, Chicago, IL.
1950 *Curing of Fishery Products*, U S Dept. of the Interior, Fish and Wildlife Service, Research Report 18, Washington, DC.

Luhr-Jensen and Sons, Inc, 1980 *Little Chief Brand Home Electric Smoker. Recipes. Operating Instructions*, Hood River, OR Luhr-Jensen & Sons, Inc (1980 edition).

Mariani, John F., 1983 *The Dictionary of American Food and Drink* Ticknor and Fields, New Haven and New York.

People of 'Ksan, 1980 *Gathering What the Great Nature Provided: Food Traditions of the Gitksan*, University of Washington Press.

Rosten, Leo, 1968 *The Joys of Yiddish*, McGraw-Hill, New York.

Schuchat, Molly, 1987 Personal communication.

Singh, Ram Raj Prasad, 1966 *Aboriginal Economic System on the Olympic Peninsula, Western Washington*, Sacramento Anthropological Society, Paper no 4, Sacramento, CA.

Smith, Marian W., 1940 *The Puyallup-Nisqually*, Columbia Univ. Press, N Y.

Steward, Hilary, 1977 *Indian Fishing: Early Methods on the Northwest Coast* University of Washington Press, Seattle, WA.

Underhill, Ruth M., 1945 *Indians of the Pacific Northwest* Sherman Institute Press, Riverside, CA.

United States Department of Commerce, Bureau of Commercial Fisheries, 1917 *A Practical Small Smokehouse for Fish*, Washington, DC.

NOTES

1. I wish to thank my wife, Kay Theodoratus, for her editorial assistance in smoothing out this paper and shortening it to a more realistic length.
2. The Pacific salmon all belong to the genus *Oncorhynchos: O. tschawutscha* (king, chinook or spring salmon), *O. nerka* (sockeye), *O. Kisutch* (coho or silver salmon), *O. gorbuscha* (humpback salmon), and *O. keta* (chum or dog salmon). *Salmo gairdneri*, the steelhead, is normally classified as a trout.
3. These booklets, circulars and leaflets are all too often overlooked by ethnologists, nutritionists and food historians. I was first made aware of their importance in Kerstin Eidlitz's *Food and Emergency Food in the Circumpolar Area*, Studia Ethnographica Upsaliensia, no 32. Uppsala, 1969.

Technical Papers

Figure 1. Production of byrghel. *The cooked and dried grain is dehulled in a stone crushing mill shaped like a mortar. All the illustrations for this paper are drawn by Tom Siwinski.*

Figure 2. Another more efficient dehulling method using a single horse gear.

TRADITIONAL METHODS OF FOOD PRESERVATION USED BY MODERN ASSYRIANS

Michael Abdalla

The most important traditional methods of food preparation used by modern Assyrians are described; but not exhaustively, since their diet is very rich and varied. A large part of the article is devoted to byrghel, after bread the most widely used wheat product in Assyria since the 9th century BC.

Modern Assyrians live in Mesopotamia, in the area which belongs to several countries such as Turkey, Syria, Iraq and Iran. In general they represent the peasant community who live in the country and support themselves from agriculture (although at present many Assyrians also live in towns and work as merchants or office workers, or—if they have university education—as engineers, teachers or doctors). The distribution of the Assyrians across the various countries has caused numerous political conflicts and as a consequence, large emigration. Nowadays, concentrations of the Assyrians are found in the Middle East, Europe and North America; large Assyrian communities also live in Scandinavia, particularly in Sweden.

The traditional Assyrian culture is a village culture and I intend to focus on it in my paper. However, I wish to emphasise that even those Assyrians who have lived in West European and American metropolises for two generations still maintain very close ties with the mother country and sacredly preserve their national tradition including the culinary tradition.

Nonetheless, it is the Assyrian peasant who still appears to be the principal guardian of the national tradition.

This presentation is based on my own observation made in my youth in a North-Eastern Syrian village and on the tape recordings of old Assyrian women living in the Tur Abdin region of Southern Turkey, near the Syrian border.

Until quite recently peasant farming was self-sufficient, and this made it necessary to store and preserve the crop until the next year's harvest. To some extent, this continues even today.

A food stock, generally called *mozuno*, is prepared in the season appropriate for a particular food source. The greatest amount of work is done in the summer while winter is the time of maximum consumption. All the members of one family, usually accompanied by a large group of relatives and neighbours, engage in the preparation of the food stock. Such a group activity appears to be the characteristic feature of Assyrian culture.

This paper describes various food preservation methods used by the contemporary Assyrian peasant both for plant and animal origin products. A brief survey of the dishes prepared from them will also be provided. Due to the climatic conditions in the regions inhabited by the Assyrians, sun-drying is a very important method of food preservation. Another characteristic feature is their tendency to preserve partially prepared 'semi-products' (milled, quern ground, cooked, preserved in sweet grape juice, etc) rather than purely raw materials (grains, fruits, etc).

COMBINED METHODS
COOKING AND SUN-DRYING. The most important product obtained by this method is *byrghel* (synonyms *gurgur, hlula*). After bread, it is the most important wheat product for the Assyrians and has been known since the 9th century BC. Typically a family assigns to the production of *byrghel* about one tenth of the wheat grain which it expects to be available for consumption.

Wheat grain (*hete, hytte*) from the recent harvest, weighing 50-150 kg, is thoroughly cleaned and poured into boiling water in a big metal kettle (*qaqwo*). The kettle, made of a thin sheet metal (0.8-1.5 m in diameter and approximately 1.2 m in height), is positioned on three stone supports. When firewood is in short supply any other inflammable materials such as dry animal excrements, rags and recently even tyres are thrown into the fire. Cooking lasts two to four hours, until the grain is fully cooked and soft (assessed by simple inspection of its appearance, taste and texture). Special measures are taken to keep the grain surface from cracking. Cooked grain is called *danoke* or *shliqa* and its nice aroma attracts the attention of the children from the neighbouring houses. Their bowls are generously filled with portions of grain by the master of the house. The children take this gift home, where it is eaten with butter and salt or sugar.

Danoke is transferred to the flat roof of a house, heaped up into a pyramid on straw mats or rugs and left overnight. The next morning the grain is spread over the whole roof surface in a uniform thin layer. After two or three days the grain is fully sun-dried (down to a moisture content of about 8%). Then, the shrunk and darkened grain is lightly sprayed with water and dehulled in a stone crushing mill shaped like a mortar (*gurno, hashulta*, Figure 1). Two men, sitting opposite each other, beat the grain with wooden hammers while a woman seated beside them mixes the grain in the *gurno* by quick hand movements after each stroke.

Another, more efficient and simpler dehulling method used in larger communities is called *dang*, which means a single horse gear (Figure 2). *Dang* consists of two large flattened and rounded stones. One with a shallow cavity, is positioned horizontally; the other, positioned vertically, is moved by a horse and rolls over the grain.[1] The hulls are winnowed away after the whole mass of the grain has dried. Dropped grain falls to the ground while the wind blows the chaff (*pyrto*) away.

Figure 3. The dehulled grain is disintegrated in the quern.

Dehulled grain is disintegrated in the quern (*gorysto*). The quern consists of two circular and flat stones (6-10 cm in thickness, 40-60 cm in diameter); the low stone is fixed in the ground while the upper stone is rotated manually around a wooden axis (Figure 3). Grain is manually fed through the hole in the centre of the quern stone. Ground grain can differ in particle size depending on the surface porosity of the quern stone; on the weight of the upper stone; and on its speed of rotation.

Next the ground grain is screened and separated into three parts, each with a different taste. Each part, combined with chick-pea (*hemṣe*), lentil (*tlawḥe*), broad bean (*baqylle*), meat (*baṣro*), milk (*halwo*), eggs (*b'e*), fresh vegetables (*yaruqutho*) or fish (*nune*), is used for the preparation of such dishes as *ballo', tyrḥayno, kutle, kifte, dawqe, tyḥrata, tabbule* and others. These are consumed throughout the year in fried meat balls, stuffed cabbage, soups and stews.

Byrghel shows no changes in taste and flavour during home storage in goatskins (*mzeethe* - 'th' as in 'this') for more than one year.

A completely different product called *basteeqe* is produced using a somewhat similar technology. The raw materials include a juice obtained from grapes (*c'enwe*) and a fine wheat flour (*qamḥo*). The grape juice is clarified by the addition of a mineral substance called *shkurto* brought from the neighbouring rocky hills. After some time the mineral sediment is removed and the clear juice is cooked. The foam that develops (*kaffe*) is collected just once. Then, while the juice is being continuously mixed, the flour is gradually added in the proportion of five parts juice to one part flour. The thickening juice, called *hawdal*, is flavoured with spices such as cinnamon, nutmeg and black pepper, mixed and poured out onto sheets of cloth. *Hawdal* is evened out by hand to a layer approximately 1 mm thick, and left to cool for several hours. Then the sheets are hung over strong strings, the coated surface facing up and left for further drying. The *basteeqe* layer is removed with a wet cloth which moistens the sheet and makes the separation easier. Its slightly wet surface is then powdered with flour and is cut into equilateral triangles (of approximately 20 cm). It is stored in clay chambers called *kore* or *esre*. *Basteeqe* is characterized by great elasticity.

In parallel to preparation of the *basteeqe*, the boiling juice, before and after the addition of flour, is used for the preparation of another improved product called *'oqoude*. Here, various fruits, such as dried figs (*tene*), soft after soaking in water for 24 hours, dried nuts (*gawze*) and almonds (*lawze*) are strung on a thread (4-6 cm apart) in the form of a long chain. The chain is dipped in the juice and in *hawdal* several times. Each time the fruits must dry so that a new layer of juice will coat the fruits and thread. The final product is a fruit chain coated with a 2 cm-thick layer of sweet, congealed grape juice. If the steps during the preparation of *basteeqe* and *'oqoude* are fully synchronized, both processes can be finished at the same time.

Haleele is a variety of *basteeqe*. For its preparation the same spices are used, but a larger quantity of flour is added (ratio 1:4) and the result is coarser, rather like macaroni semolina (*semdo*). The *hawdal* thus obtained is poured into pans to form thick layers (2-3 cm) which, after congealing (three to four hours), are cut into 5 x 5 cm rhombs or squares and left to dry for one to two days. The pieces of *haleele* are darker and crisper than the pieces of *basteeqe*, and they are stored, covered with flour, in jute sacks.

STEEPING AND SUN-DRYING. Raisins (*abshotho* - 'th' as in 'three') are very likely to be the only product obtained by steeping and sun-drying. Their preparation falls between 6 and 15 August, the dates of two important church holidays; St. Gabriel and Ascension Day, solemnly celebrated by the Assyrians. August is the hottest month of the year, popularly called *ṭybbah*, which simply means 'cook'. At this time, all the agricultural products, including grapes, are fully ripe.

In the Tur Abdin regions mentioned in this paper, seven grape varieties are known. Four of them, i.e. *qadyshiotho* ('th' as in 'three'), *byzyk*, *zaynabi* and *d'rayso*, are classified as dessert-grapes. The next two, *zaytoje* and *qarqush*, are processed into *abshotho*; and the last one, *mazrune*, is assigned for the production of syrup (*dibys*, *nepuḥta*). Some of these varieties originated in Tur Abdin, as we can tell from their names, which are unknown anywhere else. For example, the word *qadyshiotho* means 'holy', *d'rayso* - 'royal' and *zaytoye* - 'olive'.

Abshotho is always prepared in a vineyard since a large area free from intrusion by strangers is necessary. The grapes are steeped in a lye, obtained from the ash of burnt maize stems (*dahno, hatrumaye*), or from slake lime (*kalsho*) by boiling in water. This water (lye) is then decanted and transported to the vineyard in goatskins (*qarbe*) by donkeys. There, it is poured into a thick-walled metal vat (*desto*) and boiled. Freshly picked bunches of grapes are dipped in this water for a moment. After steeping, the bunches of grapes are put on mats, one close to another, in a well sun-heated, breezy, properly flattened and trodden place called *meshtohe*. The raisins are detached from stems and stalk (*qasluse*) after one to two weeks of drying and then stored in large jugs (*gdone, borynta*) in the house.

SALTING AND SUN-CONCENTRATION. Tomato paste (*maye d'bajanat, dybys d'abrushe smoqe*) is a product used almost every day in Assyrian cooking. It gives dishes both a distinct subtle taste and a red colour. Each home will have a stock of tomato paste amounting to 12-20 kg per year.

Tomato paste is prepared from a special variety, yielding big fruits (sometimes 1 kg each) with a small amount of seeds. Tomatoes usually come from a farming family's private plantation or are purchased at a market in an amount of 120-200 kg at a time. The tomatoes are sorted, washed, placed on a roof and squashed manually. The juice is then strongly salted and left on the house roof in flat aluminium pots. The water evaporates from it by the action of the sun. The thickening juice is stirred two to three times daily. Concentration is considered finished when a metal spoon put into the paste does not sink. Thus the minimum concentration of solubles is 55%.

The resulting tomato concentrate is very stable and shows no signs of spoilage on storage for a year in open glass jars. A dish (e.g. for 7 persons) with the addition of 0.5-1 spoon of the paste needs hardly any more salt. Tomato concentrate is a favourite delicacy of the children, who eat thick layers of it on flat breads abundantly powdered with dried peppermint.

SALTING AND DRYING. The amount of meat and meat products consumed by the Assyrians is not large, but meat products are expected to be present in the home all year long. Fresh meat is only eaten from time to time but preserved meat, called *qalyio*, is consumed almost every day, particularly in the spring. On the farm, one selected cow is fattened, starting in the early spring, to be slaughtered in the autumn. Boneless meat is cut into potato-size pieces, strongly coated with salt and put away for one to two days. Some of the pieces are then stewed, until tender throughout, in twice the quantity of tallow (*tarbo*) melted in a large pot over a fire. (Tenderness is evaluated on the basis of the meat's colour and structure.) The pieces of meat are, as it were, sealed in the tallow, isolated from the air and thus preserved. Cold *qalyio* is very hard and very often a spoon breaks while one is taking the product out of the pot.

Qalyio, which is very tasty and aromatic, is a sort of 'instant' food. It can be added to all meat and half-meat dishes, or to scrambled eggs (*sfero*), or can be consumed by itself on cold mornings. It only needs to be slightly warmed up before consumption.

SALTING AND SUN-DRYING. The next form of preserved meat is called *qadeedo*. It is prepared from defatted cattle meat. Pieces of meat are powdered with salt, left for three to four hours and then pressed between two flat stones for two to three days, allowing water to flow out. Pressing also flattens the meat into pieces about 2 cm thick, 20 cm long and 10 cm wide. These slices are abundantly coated with a pulp made of crushed garlic and red paprika, then hung over strings and sun-dried for about 10 days. *Qadeedo* is stored in wooden boxes in a dry place for up to two years. Before consumption it is cut into smaller slices which can be eaten without additional cooking.

Sharadin is another form of preserved meat. Fatty beef is ground, abundantly seasoned with salt and black pepper and put into small 2 x 20 x 15 cm cloth bags. The bags are sewn up and pressed with a stone for three days to allow water to flow out. The pressed meat is hung in the sun for about one week. Before consumption the product is cut into smaller pieces which can be stewed with tomatoes or added to scrambled eggs. Both *qadeedo* and *sharadin* are consumed for breakfast or supper.

SUN-DRYING AND ROASTING. The product obtained by sun-drying and roasting is called *ryshtan* and is comparable to macaroni. Wheat dough (*laysho*), without a leavener, is manually formed into a flat cake up to 2 mm thick and is cut with a knife into pieces about 1 cm x 1.5 cm. These are sun-dried, roasted in a frying pan and stored in metal containers.

CONCENTRATION BY BOILING

Crystalline sugar was until quite recently unpopular among the Assyrians and it has only been used to sweeten tea or coffee. These beverages are not traditional Assyrian beverages and became popular only after World War Two. Dishes and desserts, on the other hand, are still sweetened with grape syrup produced from *mazrune* grapes, which have a high sugar content.

Bunches of grapes, after removal of any damaged berries (*fesqe*), are placed in large cloth bags. The juice is pressed out using special equipment, which is jointly owned by several villages: it consists of two manually shaped stones called *ma'serto* (Figure 4). The upper stone, about 2 m in diameter and up to 0.5 m thick, has a vertical hole, about 5 cm in diameter, hewed out of its centre. About 20 cm from the edge of the upper surface of the stone there is a wide circular hollow (about 60 cm in diameter) with a number of grooves reaching the hole radially. The lower stone, shaped in the form of a cube about 2 x 2 x 2 m, has a chamber of about 1 m in diameter hewed out of its interior. The chamber is used to collect the juice and has a bung-hole in its bottom. There is access to the chamber through a large frontal opening so that it can be washed and prepared.

The bags containing bunches of grapes are placed in the upper stone hollow and trampled with the feet. The juice flows along the grooves and down through the vertical hole into the chamber of the lower stone. It is then collected in goatskins (*gawde*) placed at the side outlet of the chamber.

The goatskins are carried back to the farm, where the juice is poured into thick-walled metal vats and boiled. As foam forms, this is collected from the surface during the first stage of boiling. The complete process of concentrating the juice takes several hours. The density of the juice is controlled with the help of a wooden stirrer; when single drops of juice run slowly off the stirrer, the syrup is ready.

An aromatic herb called *qoshyle* by the local Assyrians is added to the juice during boiling. Grape skins, seeds and stalks are removed from the bags, dried and used as animal feed.

The syrup is stored in metal or thick-walled clay containers. During the fasting season it is an indispensable dish ingredient. Children dip bread in the syrup. The syrup is also used to prepare such dishes as *qursykat* and *dawqe*, which are similar to pancakes.

Figure 4. Special equipment for making grape juice from the grapes which are placed in large cloth bags.

SALTING

The preserving properties of table salt are utilized by the Assyrians for meat and other raw materials. After a well-fattened animal (usually a sheep) has been slaughtered (usually in September) boneless meat is cut into long and thick slices (about 10 x 5 cm) and bony meat is cut into lumps. The meat is then thoroughly salted and left in a shaded place for three to four days. After that the meat is coated in salt one more time and placed (slices and lumps separately) in wooden boxes or straw baskets, for storage in a sufficiently cold place. Salt is partly leached out in water a day before the meat is to be consumed.

Such meat, called *maksad*, is widely used for cooking purposes particularly in the winter and serves as an ingredient of such dinner dishes as *hyntyie* or *garso* (cooked wheat grain), *aprah* (stuffed grape leaves), *shamborakat* (large ravioli) and others.

SUN-DRYING

The hot and dry local climate allows the Assyrians to take advantage of the weather for food preservation by drying. A kind of acid, which can be used for food souring, is produced from grapes. For this purpose less mature grapes of a high organic acid content are picked and squashed, and the collected juice is sun-dried for one to three days. The resulting syrup, called *hysrme*, is stored in glass containers and used for food souring throughout the year. A small amount (about half a spoon) of such syrup gives food a unique taste and a light-olive colour without foreign flavours.

Sheiraye is another product obtained by sun-drying, frequently preceded by drying in shaded places. This product is very similar to vermicelli. Only women are engaged in its preparation. Hard, round pieces of dough are shaped into the form of a string up to 50cm long by rotary movements of the palms of the hand, previously dipped in vegetable oil. Then the string of dough is coiled around the left arm and small pieces of dough about 1-2 cm in length and 0.5-1 mm in diameter are gradually torn off the end of the string with the fingers of the right hand. *Sheiraye* is then sun-dried on straw trays.

In the past *sheiraye* was prepared by moonlight (*sara*); after supper and when children had gone to sleep the women gathered on the family wooden bed situated on the roof of the home. Social talk and common activities carried out there are called *shahro* by the Assyrians. Both the word *shahro* and the Assyrian name of the moon (*sara*) sound very much like the name of this product.

Towards the end of the 1950s, machines for the production of *sheiraye* were introduced in the villages (Figure 5). These circulated from house to house, hauled by oxen or pushed by a man. Such a machine consists of a container holding up to 50 kg of dough, closed with a movable piston from the top. As the piston moves down, the dough passes through the holes located in the bottom of the container. Then the dough threads are initially dried in the air stream directed from a two blade impeller fan. The fan is fixed to one of the three wooden legs of the machine and rotated manually. Partially dried dough threads are cut into 20 cm long pieces and placed on a spread cloth. They are broken into short pieces after sun-drying.

Sheiraye is added to the coarse grain *byrghel*, or to the rice, in the amount of up to 15%; this makes the dishes more tasty. The product is also consumed separately after roasting in a small amount of vegetable oil.

A local Arabian Sheik invited to our home once, enchanted by the taste of a dish with *sheiraye*, asked my father to join him on a trip to the town to buy *sheiraye* seeds which he intended to sow in his fields. It was very difficult to convince him that sheiraye is made from an ordinary wheat dough.

Dry parsley leaves, egg-plant (*bnoth gane*), tomato, pieces of melon (*fateehe*), and paprika (*felflo*) are also prepared for the winter food supply.

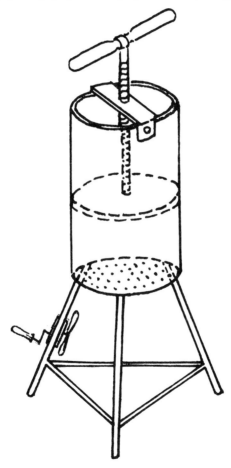

Figure 5. Special equipment for making sheiraye *introduced in the villages around 1950.*

PRESERVATION BY FERMENTATION

LACTIC FERMENTATION. The products preserved by lactic fermentation have a long tradition among the Assyrians and are very popular. Materials to be fermented include unripe tomato, radish (*fuje*), young carrot (*gezore*), cucumber (*boseene*), paprika, cauliflower (*fawhe*), cabbage (*qarambo*), eggplant, string-bean (*ḥabse*), beet (*shalghme*) and others. After the vegetables have been sorted and washed, they are placed together in glass or stoneware jars and covered with brine. Seed vegetables are punctured before processing. The brine concentration should be so great that an egg will float on the surface only half submerged. Some grapes are added to the brine. The jars with vegetables and brine are placed in the sun for several hours and then transferred to a shaded place. The vegetables are ready for consumption after 10 days and can be stored for several months.

The same preservation method is applied to young grape leaves which are picked in the early stage of growth. They are used instead of cabbage for the preparation of various dishes with stuffing similar to stuffed cabbage. Raw fermented leaves are frequently eaten by children. Stuffed cabbage made from fermented leaves has a nice sourish taste.

Another delicacy called *tyrḥayno d'maye* or water *tyrḥayno*, is prepared in order to diversify fast-day food and to make additional use of the wheat grain. Water diluted leaven is mixed with a small amount of medium size *byrghel* grains and left to ferment for two days. After the grains have been strained the remaining liquid is poured onto dehulled and overcooked wheat grain. Further fermentation in a glass jar lasts three days and the resulting dish takes the form of a cold soup.

The holiday version of *tyrḥayno* is flavoured with animal fat and the leaven is replaced with yogurt which is mixed with the wheat grain in the ratio of 2:1. Whey is often used instead of yoghurt. In that case the ingredients are boiled for two hours, until thick. The fermentation, performed in a shaded place, lasts two to three days and the resulting pulp is formed into egg-sized balls and sun-dried. The balls are soaked in water before consumption and cooked in the same way as a breakfast milk soup.

ALCOHOLIC FERMENTATION. The prohibition against drinking alcohol by Islam and the contempt shown by Moslems (particularly the ruling group) for people consuming alcoholic beverages have probably limited their consumption among the Assyrian Christians to the most important occasions only. They are served only during solemn occasions such as the birth of a child, weddings and church holidays.

In fact, only two alcoholic beverages are known: wine (*hamro*) and anise-flavoured vodka (*'araq*). Wine is produced from red grapes and (less often) from figs at the turn of August and September. Juice from the grapes is poured into a clay amphora (*g'dono*). The hole is then carefully sealed with clay to prevent air intake and 'breathing' of the fermenting juice. Amphoras are stored in straw for 40 days and then the wine is poured into new clean amphoras which are kept in a cool place, often underground, for several years.

My tape recordings indicate that the procedure of preparing *'araq* is highly original and rather complicated. *'Araq* is produced from fermented grapes (*zabile*) using quite a few variously shaped containers, with names like *dygdan, shoyefto, desto* and *lagan*. I was not able to reconstruct precisely, on the basis of the tape-recordings, the method of preparing *'araq*; nor have I ever witnessed its preparation. However, the tape recorded word *da'tho*, meaning sweat (released by the grapes, of course), is identified with alcohol. In Arabic, the language which has gradually supplanted the Assyrian language, starting from the 7th century AD, *da'tho* is *'araq*.

PRESERVATION METHODS OF SOME MILK PRODUCTS
The large number of sheep and less numerous cows and goats kept on each farm produce a large amount of milk. Milk and milk products are consumed every day. Milk fat is considered a very tasty and rich dish additive. The beestings is also a favourite delicacy among the children.

Yogurt (*Qaṭeero*) is the most popular milk product. It has a jelly-like consistency and a splendid flavour. It is prepared on a day-to-day basis and is consumed in amounts greater than fresh or pasteurized milk. But the most remarkable stable milk products are melted fat and dry curd balls called *qashka*. Every morning in the spring time fresh yogurt (innoculated the previous evening) is poured into goatskins which are then hung up in a vestibule adjacent to the western wall of the house. Regular and skillful shaking (*Myo'o*) of the yogurt (involving what might be described as a to-and-fro motion) causes the fat globules to clump. The butter (*zebdo*) obtained in this way is collected for several days and the plasma is separated from it by heating, leaving pure milk fat. The milk fat is salted and stored in clay or metal containers or in goatskins.

The remainder of the yogurt is buttermilk and is used for the preparation of another two products (see scheme). Buttermilk is heated up and the precipitating curd (*jajeq*), similar in appearance to cottage cheese, is salted and mixed with garlic leaves. *Jajeq* combined with this plant is stable and is usually stored in clay containers buried under the unhardened floor of one of the rooms. It is primarily a supper dish eaten on summer days and resembles the Polish 'cold soup' when diluted. Salted *jajeq* is frequently put into bags made of white cloth and hung up to allow the remainder of the whey to drain off. Oval or spherical shapes are then formed out of the thickened mass and dried in the sun. The product obtained in that way, called *qashka*, is used for the preparation of a popular winter dish called *gabula*. Hard *qashka* balls are mixed with water and rubbed against the wall of a special bowl made out of a fig tree trunk. Then the solution is poured onto warm, freshly cooked wheat grain placed on a dish in such a way that a hollow in the middle of it can be filled with the solution. In the end, hot fat is poured on the solution.

SCHEMA FOR PREPARATION
OF SOME MILK PRODUCTS

FRESH MILK

PASTEURIZATION
(about 6 p.m.)

COOLING AND INNOCULATING WITH YOGURT CULTURE

OVERNIGHT STORING OUTSIDE THE HOUSE

YOGURT

SHAKING UP IN GOATSKIN
(about 6 a.m.)

BUTTER

HEATING

MILK FAT   PLASMA

BUTTERMILK

HEATING

WHEY   CURD

WHEY DRAINED
OFF

BALLS FORMED
AND
SUN-DRYING

QASHKA

OTHER METHODS

The following preservation methods of raw materials and food products are used on a limited scale.

\# Keeping various cheese varieties (*gweto*), olives (*zaytune*) and egg-plants flavoured with garlic (*tumo*), and nuts in vegetable oil.

\# Storing unripe water melons and tomatoes in chaff (*tawno*) for consumption during the family feast on New Year's Eve.

\# Drying, cooking or roasting of sunflower (*sghath l'shemsho*) seeds and gourd family seeds, roasting of nuts, acorns (*balute*) and chestnuts (*kystana*).

This paper has only briefly described the most important folk methods of food preservation used by the modern Assyrians. I think that the topic has not been exhausted since the common Assyrian diet is very rich and diversified, and the majority of products come from family larders. However, some of the products mentioned, for example, *byrghel* and *tyrhayno*, are nowadays widespread in the Middle East. They are also recognized in North American and West European cuisine.

A growing number of scientific papers are devoted to these products and the techniques of preparation are regularly improved, and even in many instances patented. It is unfortunate that the patents have not been issued to the inhabitants of Assyrian villages, but to the well-equipped laboratories of highly developed countries. However, it is important to be aware of the origin of the Assyrian foods we consume, and to know that these are still being produced with the use of archaic methods by the descendants of Sargon, living in remote Turkish villages.

NOTE

1. The modern dehulling devices that have recently been brought to the villages have not gained popularity because they damage the grain and give it a spherical shape and poorer taste.

PRESERVATION TECHNIQUES TO LENGTHEN SHELF-LIFE OF BREAD

Margaret Louise Arnott

Preservation methods which have received experimental research in an effort to extend the shelf-life of bread are: low temperature storage; exposure to ionizing radiation; dehydration; heat treatment in hermetically sealed containers; and microbiological inhibition with chemicals. Of these, only freezing has achieved appreciable commercial success.

Most of us remember the bread box or bread bin on the pantry shelf where we kept the bread, and some of us still use one. If you ask most bakers the best way to preserve bread, you will get the answer: 'freeze it'. But if you ask Ray Lanci in Philadelphia, he will tell you to make it of the best protein flour and never freeze either the dough or the bread.

Another baker (Joe DeRugeriis) says that all his breads can be frozen and kept for two months, except for the French bread which should be frozen for no more than seven days, otherwise there will be a substantial loss of taste and texture. He recommends his breads be frozen in air-tight bags, preferably plastic ones. In order to restore their original taste and appearance, they should be thawed at room temperature and afterwards heated at 350°F for ten minutes. It is important to thaw bakery products before heating to insure an even distribution of heat, otherwise the outside will burn while the crumb will remain frozen.

Today bread baking and preserving is a matter of technology. All types of baked goods are marketed frozen. The market value of frozen baked goods was $557 million in 1975. Not only is bread frozen and shipped from a central bakery to supermarkets, but a large volume of frozen doughs, in the shape of bread and rolls, is being sold to the consumer for home baking, and it is also being sold to the store to be baked in the shop.

The Brown 'N Serve products use no different mixing process than fully baked breads, except that they are baked at a lower temperature for a shorter time. They must be baked long enough to gelatinize the starch and set the dough so that the cooled product will hold its shape, but not long enough to brown the crust. Normal temperatures are then used to finish the baking process, but the product must be left in the oven long enough for the heat to reach the center and at the same time to brown the crust. The advantage of half-baked bread products over the fully baked ones is that staling is unimportant. Staling of bread refers to all changes that take place after baking and can occur without loss of moisture. When the product is later baked at home, the dough is refreshed and the product has the taste and texture of a freshly baked item. However, sugar, milk, and other ingredients that hasten browning must be kept to the minimum so that the browning process will be retarded to allow time for the heat to penetrate to the frozen center. This process is satisfactory for rolls but not for loaves of bread, otherwise a half-baked crumb would result.

The use of the microwave oven is changing the way bakeries are packaging and marketing their products. Microwave ovens are unsuited for browning breads because bread dough tends to become rubbery when heated in a microwave, therefore bakers are studying ways of modifying doughs and cooking packages to protect the ingredients from the effects of microwaves. One large company is considering possible ways of creating doughs that can only be used in microwave ovens.

Freshness of bread is prolonged by the use of glycerine saturated fats, such as glycery monostearate. Bread made with oil or fat will keep longer than that made only of flour and water. Thick crusts help the interior of the bread retain its moisture, thereby extending its shelf-life.

Mold inhibitors, called antimycotics, are important in protecting the keepability of bakery goods. For bread, calcium propionate is used most frequently, while sodium salts of propionic or sorbic acids are used as mold retardants in chemically leavened baked goods. While the propionates have a marked flavor disadvantage, they have little adverse effect upon the activity of yeast which would prevent the use of sorbic acid in yeast-raised products.Other mold inhibitors sometimes used to control microbiological growth are vinegar, lactic acid, monocalcium phosphate and sodium diacatate.

The bake-at-home pizza is another bread product that has been commercialized—by a Philadelphian who supplies pizza shells that have been cooked for a few minutes and sold to be completed and cooked at home. These half-baked bread shells can be kept in the home refrigerator for two or three days before being made up into pizza and baked.

Research has shown that irradiated food reduces the risks of infection from food-borne organisms. However, in the United States, irradiated food products will not find a market until the question of suitable packaging material is settled.

Preservation techniques used for bread which have received considerable experimentation in an effort to extend the shelf-life of bread are low temperature storage, exposure to ionizing radiation, dehydration, heat treatment in hermetically sealed containers, microbiological inhibition by chemicals. Although hermetically sealed metal containers provide some protection to baked goods, particularly the fruit breads, they are not a perfect preservation method. The use of dehydration is limited to such products as melba toast and zwieback, which are not strictly speaking products of dehydration, since this term implies the possibility of reconstitution by adding water, but some research has been done by the Armed Forces in this area. Chemical inhibitors have only a small effect on shelf-life. Of the experimental efforts to extend the shelf-life of bread to the present only freezing has been a commercial success.

REFERENCES

Baily, Adrian, *The Blessings of Bread*, Paddington Press, New York, 1975.

Bakery Technology & Engineering, 2nd edn, AVI Publishing Co, Westport, Connecticut, 1972.

Encyclopedia of Chemical Technology, Vol 3, A. Wiley - Interscience Publication, New York, 1978.

Etter, Gerald, 'The Breaking of Bread from Its Side Role' in the *Philadelphia Inquirer*, August 24, 1986, H 1-2.

Fish, Larry, 'Made to Go' in the *Philadelphia Inquirer*, May 5,1987, C 1-2.

Pomeranz, Yeshajahu & J.A. Shellenberger, *Bread Science and Technology*, AVI Publishing Co, Westport, Connecticut, 1972.

Spicer, Arnold (ed), *Bread: Social, Nutritional & Agricultural Aspects of Wheaten Bread*, Applied Science Publishers, London, 1975.

Wolf, Ron, 'U.S. Stymied on Food Sterilization by Irradiation' in the *Philadelphia Inquirer*, April 20, 1987.

'Warming to a Microwave Crowd' in the *Philadelphia Inquirer*, April 1, 1987.

SEQUENCES IN THE CONSERVATION OF MILK PRODUCTS IN SCOTLAND

A. Fenton

The sequences of milk conservation described in this paper have arisen from a combination of shrewd observation and the pressures of circumstance. They reflect necessity within self-contained communities and, on a more general level, the importance of milk to the national coffers.

Milk is a precious and age-old food, but whatever its source, it is a perishable commodity. Nowadays, there is a plentiful and apparently unlimited supply of milk that is alleged to be fresh, though coming from the pooled produce of a multitude of cows and pasteurised out of all recognition. In earlier days, there were no scientific aids for extending the life of this valuable liquid, and in the course of nature, the flow from the cows or other stock diminished and dried up before the appearance of new progeny, and during the winter. The further processing of raw milk into forms that allowed for longer keeping and availability during barren times, was, therefore, an early necessity.

Of these forms, butter and cheese are the most common, though by no means the only ones. From a historical viewpoint, such products have to be related to the nature of their sources. The milking of sheep, still common in many parts of the world, had effectively come to an end in Scotland by the late 18th century. Before then, their milk was widely used for human consumption, along with that of goats and cows, and in the 16th century rentals were paid in part in ewes' butter and cheese, especially in the more pastoral areas. The view of a writer on farming practices near Edinburgh in 1666, however, was that cows' milk was best for butter, and ewes' milk for cheese. Ewes' milk butter was evidently of secondary quality. It was widely used as grease and was often mixed with tar to make a salve for smearing sheep as a means of keeping their wool free from vermin.

Ewe milking survived till about 1800 in the hills of the Scottish Borders, an important sheep-producing area where farmers used much ewes' milk butter for sheep-salve, and also made cheese for sale to the shops in the towns. Sheeps' milk had a commercial interest for the Lowland producers; in the Highlands, on the other hand, it remained as a subsistence item for half a century more. Even after landowners had cleared people from great tracts of land in the Highlands to make way for large grazing farms for commercial breeds of sheep, it was not uncommon for Blackface sheep to be driven into pens once or twice after the lambs had been weaned, to get their milk and keep the individual households supplied with cheese. Goats also went on being used for milk and cheese in the Highland areas. Such points mark differences between the farming Lowlands and the pastoral Highlands where the knock-on effects of capitalistic attitudes to the resources of the countryside came half a century later than in the Lowands.

As long as the milking of ewes and goats went on regularly, their milk was often made into cheese, separately, or mixed together, or sometimes with the addition of warmed cows' milk. The story of butter and cheese-making in Scotland, however, is one of increasing dependence on cows' milk alone after the early 1800s. Of course, cows were also the leading suppliers of milk before that date.

Cows' milk products can be examined at two levels: official and domestic. Butter and cheese often appear in early charters and rentals, and in grants and regulations of the Parliament of Scotland from the time of David I in 1147 onwards. James VI forbade the export of cheese in 1573. Charles II, in 1661, required 2 oz (57 g) of bullion to be brought to the royal mint for each 5 cwt (254 kg) of cheese exported. Such trade restrictions imply that Scotland was not fully self-sufficient in milk products, at least in the main centres of demand. This is made clear, too, by the fact that the better-off members of society were importing quantities of English cheese from Cheshire and Gloucestershire for their private tables. There was also much internal movement of cheese and butter, by sea or in panniers on horseback, to satisfy the needs of town-dwellers and bakers' shops in towns. There was a cheese warehouse in Edinburgh in 1790; cheese and butter were for sale on the market stands; and by the early 1500s the recognised trade of 'cheesemen' was in existence. There were also 'buttermen', and buttermarkets were well-recognised institutions where butter was sold fresh or, more often, in salted or 'powdered' form.

The scale of demand, and therefore the degree of importance of these products in the dietary system, was high. On the Aberdeenshire estate of Monymusk in the 1700s, 3 lb (1.4 kg) of cheese a day were eaten, or 1095 lb (497 kg) a year. Butter was consumed at an even higher rate in the summer months, when it was fresh and plentiful. The main developmental factor, though, was urban demand, which encouraged surrounding farmers to increase the amount and improve the quality of their products. The South-West of Scotland, where the famous Ayrshire breed of dairy cows evolved, became Scotland's leading cheese-producing area and already by 1812 a farmer in Caithness in the far north was making Dunlop cheeses, named after the Ayrshire town where they began to be made, using Aryshire cows and employing Ayrshire dairywomen. His cheese then went by sea to Edinburgh for sale. Clearly, widespread trade in what was not only a conserved but a recognised quality product, was already well established, and imports from England became less necessary than in earlier times. The better off stopped turning up their noses at local cheeses.

The conservation of butter required a lot of space. Containers of all kinds were needed for long-term transport and storage: tubs, barrels, boxes and chests, and earthenware containers. Homes, cheese lofts and butter rooms in bigger establishments, and army and navy provision stores were full of them. They were the butter mountains of pre-EEC days.

Cheese was made of broken up and lightly salted curds, though salt was not always easy to get. The little ewes' milk cheeses made in the island of St. Kilda, and the goats' milk cheeses from the island of Jura were sometimes cured with ash from burned seaweed, and barley-straw ash could also be used. It appears to have been a normal practice in the Highlands to make cheeses without salt, though they were salted for a few days in a wooden keg, after they had been pressed.

Soft cheeses, including the kind known as 'crowdie', were eaten at once. They were not made with the intention of long-term keeping. They were hung in a cloth to drip and to dry in the air, and then they were consumed. For cheese to be kept longer, it was essential to press out all the whey by means of a cheese-press operated by weights, screws and the like. Some of the pre-18th century type, of long planks wedged at one end and weighted at the other, could produce cheeses of up to 19 kg in the 1790s and some of the sheep-farmers of South-East Scotland were making 1270 to 1905 kg for the market each year. Screw-operated, fixed cheese presses of stone and of iron first appeared in the dairying South-West about 1800 and rapidly spread to other parts of the country on farms of medium size or over. Smaller places made do with portable wooden presses. Home cheese-making continued widely until the First World War, when large creameries using pooled milk supplies were set up, and home-processing slowly became redundant. Grocers' travelling vans aided the process by distributing not only creamery cheese, but also cheese imported from abroad, to all parts of the country. Home-made cheese was of great importance to the diet in the past; it is likely that in spite of creameries and imports, less cheese is eaten in Scotland now, on an average, than at any time over the last 500 to 600 years, though the variety of available types is infinitely greater.

Necessity, it is said, is the mother of invention. Even fresh milk or, failing that, the whey from butter-making could be processed to make it go farther. This was perhaps less a question of preservation than of aeration, however. It was done with an instrument called a 'froth stick'. The milk was first boiled, then this stick, with a cross-head round which horse-hair was bound, was put into it and twirled rapidly between the palms of the hands. The result was 'worked' milk, with a strong froth. In the 1690s, it was said that frothing up was done five or six times, and the froth supped off the top with spoons each time. Frothed milk could be a hunger food, or could be made as a pleasant drink. Sour, thick milk below cream could be worked in the same way, as could a mixture of cream and whey. The evidence shows that frothing was practised in both Highland and Lowland areas. Such a wide distribution points to a practice of some antiquity.

When a cow had newly calved, the first—and more rarely the second—milking was put into a dish, flavoured with salt and sugar, and warmed in the oven till it set to a custard-like consistency. The dish, called 'calfie's cheese', does not appear to be very old, nor is this really a preservation technique.

Much more traditional, however, was a practice of long standing, well-known in Orkney and Shetland. This made use of buttermilk. When hot water was added to it after churning, this produced a white cheese-like substance. The serum that was left, after this was removed, was known as 'bland'. It was drunk fresh, or could be kept in barrels. After a time bland fermented to a sparkling stage, when it made a refreshing drink.

Bland was for long a universal drink, serving even as a substitute for ale or beer. Its keeping qualities could be enhanced by adding fresh serum at regular intervals, so that it stayed sparkling and did not become flat and vinegary. There was an element of social differentiation in its use: poorer people drank it hot from the churn, whilst the better off let it mature in casks in their stores, keeping a good supply for the future. Fishermen, who in the 18th and 19th centuries went far out to sea to fish in open boats, normally took a keg of bland with them as their drink. On the farms of the Northern Isles, bland served to augment the feed given to calves in the byres. They were not allowed to suck the cow, but were first given a drink of her milk, and then a drink of bland to make up the quantity for the calf, leaving milk enough for household use.

The white substance produced by heating the buttermilk was sometimes called 'hard milk'. It could be eaten with fresh milk, or hung up in a cloth and left to drip for a time. The outcome was a kind of soft cheese. A parallel to this was 'hung-milk', made from full milk that had coagulated due to the heat of the weather. It was put in a finely-woven linen bag and hung up till the whey had dripped from it, leaving a soft, cream cheese.

Although coagulation could take place in the course of nature, it was sometimes artificially induced. The method was to add sour milk or buttermilk to sweet milk, and then to heat it. The product was called *ost-milk* or *eusteen* in Shetland. In effect, it was a form of curds and whey. There was also a dish called *strubba*, made of coagulated milk whipped up to a creamy consistency. This was eaten with rhubarb or with pudding.

Butter is itself a result of a thickening or souring process. Milk left to stand for two or three days thickened naturally into what was known as 'sour milk' or 'run milk', and this is what was churned.

Similarly, cheese-making began with coagulation induced by rennet plus heating. A variety of substances could be used as rennet to curdle the milk or to *yirn* it, to use the common Scottish term, which is, in fact, a metathetic form of the old English *rinnan*, to run. People used, as a form of rennet, the stomach of a calf, lamb, hare, deer or pig, the gizzard of a fowl, or the leaves of plants like butterwort, *Pinguicula vulgaris*, and autumn crowfoot, of the genus *Ranunculus*. The maw or stomach of an unweaned calf was regularly kept for this purpose.

A dish that was very wasteful of new milk, but no doubt gastronomically satisfying for all that, was called 'clocks'. The milk was boiled for hours until it became thick, brown and clotted. Several gallons of milk were needed to make a moderate amount of clocks.

There was a great deal of variety in the uses of milk. The liquid could be drunk fresh, and in this form could be frothed up to make it go farther.

There were means of utilising it in naturally curdled form, and it could be artificially curdled, especially in order to produce cheese. After the churning of somewhat soured and thickened milk to make butter the remaining buttermilk could be split into a soft cheese-like substance and a clear serum that served in fresh or fermented form as a widely used drink. Whey from cheese could also be so used, and in the cheese-making South-West, it was also an excellent food for the pig-rearing industry associated with the dairying districts.

A great deal of human ingenuity has gone into working out such sequences in the conservation of milk products, no doubt resultintg from shrewd observation combined with the pressures of circumstances. The general pattern is double. At the level of the old self-contained communities, such ingenuity was a necessity to extend the period during which this most basic of all foods could be available. At the level of trade based on milk products assembled through the payment of rents and other dues, milk and milk products also had their degree of importance to the national coffers, even if not at the same level as the grain and cattle trades.

BIBLIOGRAPHY

The main sources for this study are:

W. Aiton, *A Treatise on the Dairy Breed of Cows and Dairy Husbandry* Edinburgh 1825.

A. Fenton, *Scottish Country Life*, Edinburgh 1976 (reprint 1977), 147-58, with detailed references.

A. Fenton, *The Northern Isles: Orkney and Shetland* Edinburgh 1978, 438-43, with detailed references.

In addition, *The Scottish National Dictionary* and *The Dictionary of the Older Scottish Tongue*, provided pointers to a wide range of sources.

OLIVES—AN ANCESTRAL TRADITION FACES THE FUTURE

Lourdes March and Alicia Rios

The techniques currently being used to preserve olives reveal the continuing existence of archaic and vernacular practices. They also bear the impressions left by the cultures of the various peoples who have successively occupied Iberian soil. The ways in which the olives are treated, the types of containers used, and even the flavours, are today almost the same as in former times. Olives however are also a modern, highly industrialised product, with a new function both in the diet and on the table.

Olives can be regarded as the 'seeds' of culinary traditions, the source of the symbolic values which abound in gastronomic nostalgia. They represent both the past and the present. The preparatory treatments which they require to make them edible are equivalent to a cooking process.

THE OLIVE IN HISTORY

The olive tree, according to legend, sprouted out of the earth when Pallas Athene, goddess of wisdom, hurled her spear to the ground. Since then it has become a symbol of peace, victory and life. The champions of the Olympic Games were crowned with olive branches and were anointed with *elaion* - the oil extracted from the olive tree. The wood of the tree was carved into statues of the gods, sceptres for the kings and battle weapons for the heroes.

The olive tree, *Olea europea sativa*, can be found throughout the Mediterranean basin. The characteristics of the soil, climate and humidity are ideal for its growth, and its cultivation is limited only by cold weather as the tree can barely survive temperatures lower than -12°C. On the other hand, it can stand exceptionally bad droughts and the strongest winds. It grows on land up to an altitude of 400 metres if situated on sheltered, south facing terraces.

It is found growing wild in Spain, Portugal, North Africa, Sicily, Crimea, Caucasia, Armenia and Syria. The olive is an evergreen tree and its shape and size vary according to the fertility of the soil, the climatic conditions and its exposure to the elements. It has a remarkable ability to adapt itself to different soil structures as its roots grow down very deeply into the ground, enabling the tree to obtain all the elements necessary for growth and also allowing it to resist severe droughts.

Its trunk is smooth and grey when the tree is young but becomes gnarled, wrinkled and cracked when it ages. It has persistent leaves with a life span of about three years.

The tree has a slow growth rate; even in the most favourable conditions it does not bear fruit for the first five years and takes 20 years to reach full maturity.

The olive was probably first cultivated in the coastal areas of Syria and Anatolia, later spreading to Greece. Various different hypotheses regarding the origin of the olive have been put forward on etymological grounds. The vulgar names for the tree in the languages used in the Mediterranean basin are derived from two separate sources: the Greek word *elaia* and the Hebrew *zait* became *olea* in Latin and *zaitum* in Arabic. *Olea* could be a reference to the territory to the north of Mount Olympus and *zait* is thought to refer to Said, a place in Western Egypt in the Nile Delta.

From the 16th century BC onwards, the Phoenicians spread the cultivation of the olive across the Mediterranean, first to Libya and Carthage and then on to other countries bordering on this sea.

In the far distant past the Assyrians, Greeks and Romans prepared and consumed olives with a variety of different coatings, such as honey and vinegar. *Acronyt*, the main Greek meal, consisted of bread soaked in oil and wine accompanied by olives and meat or fish. Difilo, a famous doctor born in the 3rd century BC in one of the Cyclades Islands, said that olives had astringent properties which stimulated the appetite and eased the digestive process.

The Romans were familiar with a great many different varieties of olive, some of which were harvested while still green, some when semi-ripe and others when very ripe, according to the type and what they were to be used for. The first appetizers were eaten in Roman banquets after toasting the glory of the Emperor and almost always consisted of olives imported from North Africa. Roman gastronomy had achieved such a high degree of refinement that an experienced gourmet could, we are told, tell with his eyes shut whether the olive he was eating had been picked with bare hands or whether the person had been wearing gloves. In the latter case, he could even tell what kind of leather the gloves were made of!

THE OLIVE TRADITION

The consumption of the fruit of the olive tree goes back to ancient times and this tradition has been handed down from generation to generation, from populace to populace. Over the years a vast amount of knowledge has been accumulated with regard to the treatment and preservation of olives, ranging from the simplest domestic procedures to the most advanced industrial techniques.

If we undertake a comparative study based on the cultural and linguistic influences left behind by the Roman/Arabic presence in the Iberian Peninsula, we can see how two large, clearly defined geographical areas emerge with respect to the history of the olive. In the southern and central regions there is a predomination of Arabic words like *aceituna*, *aceite* and *almazara* (olive, oil and oil mill respectively), while all through the Levante area and in Catalonia and Aragon the words *oliva*, *oli* and *molino* are used, all derived from Latin.

These differences are also reflected in the eating habits and culinary customs to be found today, illustrating the continued existence of traces of the cultures of the different peoples who have occupied Iberian soil.

The olive is enmeshed in the history of the Mediterranean basin and references to it can be found in both profane and sacred documents, showing its cultural and economic importance. The olive harvest together with the subsequent oil-producing processshave traditionally always been accompanied by festivities, and nowadays many olive growing regions hold ceremonies at harvesting time.

Numerous songs, rhymes and proverbs are based on the olive tree and its fruit, with references to the tasks of the harvest mingling with stories about the lives and loves of the country people.

CHARACTERISTICS AND TREATMENT OF THE OLIVE

Although the fruit of the olive tree bears morphological resemblances to other fruit with stones, its chemical composition and organoleptic properties display some fundamental differences: the sugar content in its mesocarp is much lower than average; it contains a bitter glucoside called oleuropeina (only found in olives); and there is a high oil content in its pulp, varying from 17% to 30%, depending on the variety.

The oil content in the varieties of olive which are intended to be eaten rather than made into oil must be as low as possible. The main varieties of this type in Spain are Manzanilla, Hojiblanca and Gordal, although there are about 15 other kinds.

According to the degree of ripeness, olives change colour from green to purple and then to black. The ripening period usually starts in September-October and finishes in January-February, depending on the individual area and variety. The characteristics of the fruit and the chemical composition

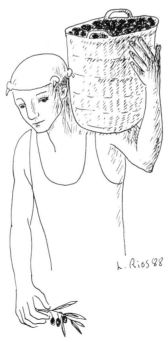

L. Rios 88

of the pulp depend on various factors: the variety of the fruit, the geographical situation of the trees, the quality of the soil, the type of cultivation, the rainfall or irrigation conditions, and finally the degree of ripeness when harvested.

The harvesting of green olives—called *de verdeo*—must be carried out with the utmost care; that is to say, when the olives have grown as large as possible before beginning to change colour - called *envero*. The picking has to be done by hand, a process called *ordeno*, to minimize the bruising caused by bashing or bumping; these have an adverse effect on the various treatments which the olive undergoes later on.

All the olive growing areas in Spain still maintain the custom of preparing and preserving olives so that they are available for consumption all year round. The people in these regions pick the olives from their own trees, if possible; otherwise they buy them. They then apply the traditional methods of preparation, using the same formulas as their ancestors—the same herbs and ingredients for the seasoning, paying careful attention to the appropriate treatment and soaking times for each variety, and using the same kind of receptacle for the maceration process, usually earthenware vats. These containers are especially suitable for the fermentation and preservation processes due to their porosity and ability to maintain a constant temperature.

The most admirable and surprising feature of all this inherited wisdom is that the various preparation and preservation processes are carried out with no 'technical' knowledge of the chemical phenomena which take place during them. Nevertheless, the know-how acquired from their parents enables the country folk to recognize when the olive has reached the appropriate stage at any point in the proceedings, whether the process to remove the bitterness, the fermentation period or the seasoning procedure. They can also calculate the proportions of the ingredients without having to measure them, ending up with top quality olives which taste absolutely exquisite.

A glass of wine and a dish of 'home-made' olives have traditionally been the pride of country people and were always offered to a visitor on arrival at one of their houses.

The phases of the moon are no longer taken into account when deciding the dates of the olive harvest, but traditionally the picking was always done in the last quarter.

The women prepare the ground around the olive trees beforehand by spreading out nets and then the men climb up ladders with a kind of basket (*macaco*) hanging around their necks. They pick the fruit off the branches and the women collect the ones that fall onto the nets. The olives are packed into crates and transported either by motor vehicle or animal traction. After being left for a period of 24 hours, they are in the optimum condition to undergo the preparation process.

The olive is made edible by means of a process which eliminates the bitter glucoside. This can be achieved by two methods: either quickly, using an alkaline treatment; or slowly, by repeatedly rinsing the olives in

water, immersing them in brine, or, if they are very ripe (black olives), putting them directly into dry salt.

Although industry has reached a high level of technology, the procedures used are based on traditional methods, using much more advanced equipment, of course.

The fast process to eliminate the bitter glucoside traditionally consists of applying an alkaline treatment, using ash or bleach. This first phase takes several hours (between 5 and 10), depending on the size and ripeness of the fruit. This has more or less the same effect as a cooking process. Afterwards the olives are rinsed several times in water.

On an industrial basis this process is carried out by immersing the fruit in soda solutions with a gravity of 1°to 5° on the Baume scale and then rinsing them thoroughly in pure water. Green olives are protected from exposure to the air to prevent them turning black as a result of oxidation. This problem naturally does not arise in the case of black olives which can be exposed to the air with no adverse effects. The fruit is then put into a flavoured or unflavoured brine for preservation.

The action of the alkaline treatment on the olives is very complex and brings about substantial changes in the fruit, including:
* The elimination, as we have already mentioned, of the bitter glucoside (oleuropeina) and other related components.
* The transformation of sugars, colouring substances, etc and a substantial increase in the fruit's permeability.

Another method of accelerating the elimination of the bitter glucoside, used in both domestic and industrial processes, consists of splitting the olives and then striking them with a wooden mallet or a stone. This

should be done on a wooden surface or a brick to prevent the fruit from slipping. The stone of the olive should not be broken. Sometimes the fruit are cut lengthways (in industrial procedures this operation is carried out by very fast machines) and placed in a large receptacle with enough water to cover them. The water is then changed on a daily basis for a period of 8 to 10 days. This operation to remove the bitter taste is traditionally known as either the 'sweetening process' or the removal of the *alpechin* (as the liquid which seeps out of olives is called). Tap water is usually used, but in the villages well or rain water which contains no chloride is preferred.

Following the alkaline treatment, if the acceleration process is being effected, or directly if the traditional formula is being followed, the olives are put through a natural lactic fermentation process. This is done by immersing them in a brine with the appropriate salt concentration, which traditionally is measured by dissolving coarse salt in water and testing the concentration by placing an egg in the mixture; when it floats thin side up there is enough salt. In industrial processes the brine usually contains a 10° or 12° salt concentration.

The lactic fermentation consists of the breakdown of sugars, especially glucose, to produce lactic acid. This transformation is brought about by means of the action of certain bacteria called lactobacilli or lactic bacteria. It should be pointed out that the fermentation process can only begin and keep going at a termperature close to 27°C. In the case of natural black olives the lactobacilli are not always present but other microorganisms such as yeasts, are developed.

During the fermentation an osmotic process takes place in which the olive yields carbohydrates and absorbs sodium chloride. The brine exercises various functions in this process. In the first phase, which lasts two or three days, part of the cellular liquid of the olives is extracted and then transformed into a culture medium which is suitable for the development of the microorganisms of the fermentation.

In the second phase, which lasts from 10 to 15 days, the lactobacilli develop very quickly and the acidification process becomes more intensified. The third phase continues until acid production stops and the medium is established. As a result of the acid produced during the fermentation process, the pH changes from alkaline (higher than 7) to acid (pH4), thereby achieving the necessary conditions to preserve the olives until the time comes to pack and market them. In some industrial preparations, and even more in domestic formulas, the maintenance brine is accompanied by aromatic herbs, lemon, vinegar, garlic and other ingredients. This liquid is called the *adobo* or *aliño*—marinade or seasoning—and in every region of Spain there are various traditional ways of preparing it. These local secrets are passed down from generation to generation, thereby maintaining the individual customs of each region. The total duration of the process tends to vary since some varieties of olive

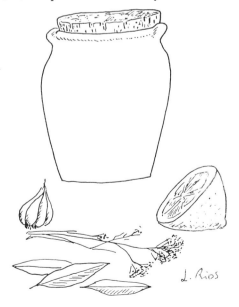

J. Ríos

can be consumed within a period of 10 days while others require a period varying between 2 and 9 months before achieving the acceptable organoleptic properties.

Black olives, which are picked when fully ripe, are washed and preserved straightaway in brine or dry salt.

Olives are sometimes sun-dried and then macerated in oil. Home-preserved olives are kept in the containers with the brine used to prepare them and are spooned out to be eaten as required with a perforated wooden ladle. Olives which are processed industrially are kept immersed in brine in large tanks until the time comes for them to be packed, when the brine is changed for a lighter solution. Depending on the variety, the olives then go through a sterilization or pasteurization process, or in some cases are simply vacuum packed in cans or jars.

THE PRACTICAL FUNCTION OF PRESERVED OLIVES
The olive, after its special preparation and preservation, is in optimum condition, both nutritionally and practically, and ready to eat.

The finished product contains all the main amino acids. There is a high concentration of leucine, aspartic acid and glutamic acid. There are considerable quantities of vitamins A and C, carotene and thiamine, as well as mineral elements, calcium and magnesium in particular.

The calorific value is quite variable and depends on the lipids contained in the different varieties.

There is also a balanced proportion of fibre, which is of great importance in the efficient functioning of the digestive system.

Besides having an anti-cholesterol value, the olive and its oil are purgative agents; that is, they eliminate the obstructions, known as lithiases, which sometimes appear in the liver and the circulatory system. They also soften the mucous membrane of the stomach and the intestine, and help maintain the glandular and sexual balance.

From a nutritional point of view, the olive is one of the cheapest foods available since, apart from its high vitamin content, it comprises trace elements and nitrogen and lipid complexes. It is therefore an ideal basic element in the diets of sportsmen, convalescents, people who are suffering from depression or are run down, old people and growing adolescents, as well as people whose skin is dull and starting to wrinkle, and whose hair has lost its shine.

The practical function of the olive lies in its small size and high nutritional value. They are easy to transport and can be adapted to the requirements of different forms of consumption, in line with the changes in the consumers' habits. These advantages could be outlined as follows:
No preparation is required.
They can be eaten on their own as an appetizer, or accompanied by a drink, such as beer, sherry or vermouth.
They can be used either as decoration or as a basic ingredient in a wide variety of recipes: pizzas, rice dishes, sauces, dressings, pates, with eggs, fish or meat.

\#   They are a basic ingredient in salads, the consumption of which has developed spectacularly in recent years.

\#   They have great potential for use in the supply industry, restaurants, etc, and are especially suitable in the preparation of meals for airlines.

\#   They are very useful in the food transformation industry as an additive or flavouring, for example in cheeses, biscuits and sausages.

\#   Their consumption is closely related to the 'Mediterranean diet' which is currently enjoying widespread popularity in the more developed countries.

The olive combines the four basic tastes: acidic, sweet, savoury and bitter, and in general terms its special characteristics allow its organoleptic properties to be well appreciated by the consumer.

The olive tree has always been closely linked to the history and civilization of the Mediterranean, and still offers us its magical fruit which, in its long journey from ancient times to the future, has evolved from being a basic rural food to its present position as an indispensable part of contemporary diets, rising up from the roots of our feeling of identity to become a nostalgic symbol of our childhood memories.

ON THE DOCUMENTATION OF FOOD CONSERVATION

Astri Riddervold

In order to understand why this and that are done when food is prepared for conservation, it is necessary to know the conserving factors which prevent the microbiological and chemical decomposition of food. When these are known, it is possible to give an accurate and thorough description of the conserving process. This paper will give a short theory and several examples.

It is tempting to state that the ability to store food is the basic factor in building a society of even the simplest kind.

Historical research tells us how food was stored as far back as several thousand years BC. Drying and smoking have been practised from the time when man first made fire his servant; salted, dried fish was eaten in Mesopotamia, Egypt and the Indus valley nearly 6000 years ago. Wallpaintings from ancient Egypt show the salting of geese in large jars. In classical antiquity, salt was used in Greece to conserve fruit and vegetables—it is recorded, for example, that salted olives were eaten at the battle of Marathon. The Romans used a whole variety of techniques to conserve both animal food and vegetables. The cookery book attributed to Apicius, *De re coquinaria*, gives fascinating recipes in this category.

In China, cabbage was preserved by fermentation thousands of years ago by methods like those used in Germany and Eastern Europe for making sauerkraut today; and cheese was made by fermented soybeans as far back as 2000 BC. The Chinese used a method famous in modern biotechnology: a certain mould living on the soybean produces enzymes which decompose the protein.

Fermented fish can be traced back to 1000 BC in Egypt; and the tradition is still very much alive in South-East Asian countries as it is in Norway, Sweden and Iceland.

Traditional conservation used three categories of agent, usually in combination:
1. natural factors like wind and temperature;
2. conserving media like salt, smoke and acid;
3. biological factors—micro-organisms and enzymes.

The only conservation technique to which a person's name and a date can be attached is bottling, invented by the Frenchman Appert in 1809. All the others have been part of man's daily life for thousands of years, as they still are.

THE IMPORTANCE OF MICRO-ORGANISMS

All food will perish, sooner or later, if nothing is done to preserve it. The decomposition of all organic material is part of the eternal recirculation in Nature. All organic material produced by Nature is after some time transferred back to Nature. Millions and millions of micro-organisms take part in this process. Micro-organisms are said to be the most important entities on earth since, without them, no other form of life could exist. In the process of breaking down organic material the micro-organisms help to start chemical processes which in the end transform all organic material to the simple components that plants can use in their growth.

Man discovered that not all these processes made foodstuffs inedible; also that some could be stopped, and others controlled. The Egyptians used this knowledge and embalmed their Pharaohs—Nature did not get them back.

So man was able to develop a whole lot of techniques for conserving their food. They learned how to prevent or halt the unwanted microbiological activity and chemical processes; and, on the other hand, to create a favourable environment for those activities which were beneficial.

DEFINITIONS
A Danish book on food conservation used for the education of veterinarians, gives these, suitably ethnological, definitions:
CONSERVATION. *A complex of different techniques, all aiming at preparing food in such a way that, when stored, it will not decompose to become a health hazard.*
EDIBLE FOOD. *A foodstuff can be termed edible if it is acceptable to a substantial number of people who are acquainted with that type of food, who know the process of preparation, and who are in a position to evaluate its condition.*

DOCUMENTATION OF CONSERVATION TECHNIQUES
When we are doing research on food conservation and trying to document the different techniques, our description of them will have little or no scientific value unless it includes full details, as close to reality as possible. We cannot rely on our informants to give us the full description; little details of great importance are often left out just because their importance is not realised. Our task is to put forward significant questions. We can, however, only ask the questions which we ourselves know to be of importance. It is not enough to know how things are done. When we deal with food conservation, we must also know why, to be sure we have answers to all questions of any importance for a thorough documentation of the technique. Our aim must be to describe the process in such a way that it can be repeated successfully.

THE THEORETICAL ASPECT OF CONSERVATION
In order to know why things are done, we must return to the micro-organisms and their ways of life. There are four types of spoilage agent which, in different ways, decompose food. Three are micro-organisms: moulds, yeasts, bacteria. The fourth group are the enzymes. It is their activities which must be inhibited in order to protect the food. Sometimes, however, their activities can be used to further the preservation and desirable transformation of the food. So it makes sense to

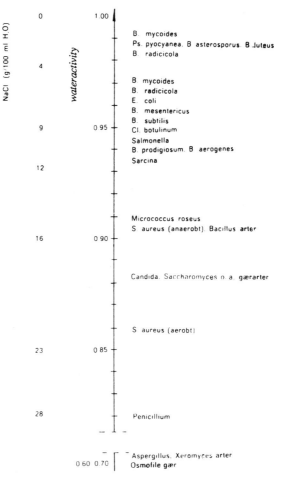

Figure 1. The lowest wateractivity where different micro-organisms can grow when all other conditions are favourable (T and pH). To the left the concentrations of salt are given which correspond to the different degrees of wateractivity (from Industriel levnedsmiddelkonservering).

provide these helpful micro-organisms with an environment which enables them to perform their good offices, but without becoming over-active.

In these four groups there are numerous species, each of which has its own particular and immutable relationship to four environmental factors. These are:

1. the amount of free (unbound) water molecules in the food, so-called water activity (A_w) or in other words the degree of dryness;
2. the temperature (T);
3. the pH, the degree of acidity/alkalinity;
4. the amount of oxygen present (O)—aerobic or anaerobic environment.

The immutability of these specific relationships is the key to every preserving process. It was only recently that scientists became able to measure the exact values of these relationships. Empirical experience, however, is hundreds, even thousands of years old, carefully handed down from one generation to the next.

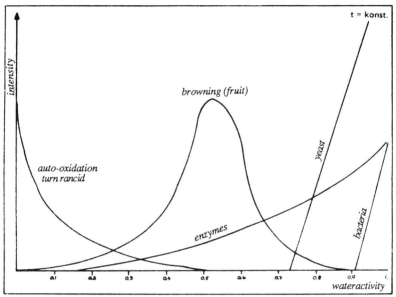

Figure 2. The intensity of different types of reactions as a function of the wateractivity when T is constant (Heiss and Eicher 1971).

WATERACTIVITY (^{A}w). Wateractivity was traditionally diminished in different ways:

1. Mechanically, by pressing the water out of the item in different ways such as beating with stones, weighting with stones or some other heavy item, or just jumping on it.
2. Drying the food by air, wind and/or heat.
3. Adding salt and/or sugar which binds the water molecules and has the same conserving effect as drying.
4. Smoking, which dries up the surface.
5. Freezing, which binds the water molecules in frozen crystals.

To weigh down the food when salt is added accelerates the drainage of the food and thus also accelerates the penetration of the salt into it. This is done particularly in areas where the temperature may be unfavourable (more than 10°C), or when for other reasons extra protection is necessary. An example is the Norwegian *gravlaks*, the 'buried salmon'. Here a very weakly salted product is wanted, and the fillet of fish, already sprinkled with a mixture of salt and sugar, is weighed down for the first two days by putting a carving board on top of it and a weight of about one kilo on top of the board.

Another example is the traditional *jamon serrano*, the cured mountain ham from the areas near Madrid. Here the piece of food is big, the temperature unfavourable, and a quick penetration of salt is essential. The fresh meat is put into a clean cloth sack filled with crystalline salt, which is placed on the basement floor. Everyone visiting the house has to go down into the basement and jump on it. The Bulgarians have the same traditions, only here the meat is put on the threshold (see Radeva).

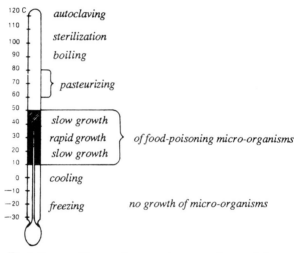

Figure 3. The influence of temperature on micro-organisms and the temperature at certain ordinary methods of conservation (Industriel levnedsmiddelkonservering, kap 19).

TEMPERATURE (T). Several methods were used traditionally to secure a favourable low temperature:
1. digging the food down in the snow;
2. putting it in a firkin hanging in a well of cold spring water;
3. digging it down in the bog; or
4. keeping it wrapped in a wet substance (a method used particularly for transport—when the water evaporates from the surface, the temperature is kept low inside).

DEGREE OF ACIDITY/ALKALINITY (PH). Some types of food naturally contain such a high concentration of acids that they are self-conserving. Cloudberries and lingonberries are good examples. But the pH of most foods needs to be lowered, and this has traditionally been achieved in several ways:
1. by creating favourable conditions for acid-producing microorganisms (in preparing sauerkraut, for example);
2. by storing the food in a sour medium, vinegar, sour milk, etc;
3. by smoking: smoke consists of a lot of different chemicals, amongst them several acids which lower pH on the surface of the food being smoked.

AEROBIC OR ANAEROBIC ENVIRONMENT (O). The fourth factor to take into consideration is the presence or lack of oxygen in the environment. Some micro-organisms (including the two most dangerous ones) live only in a vacuum, but the rest must have oxygen to be able to live. To store food in a vacuum was traditionally done in different ways:
1. in brine or other liquid, alkaline or sour;
2. dug down in the sand on the shore, in a bog, or in grain;
3. under a cover of hard fat, previously sterilized, or covered with soft goose-fat or lard, oil or butter;
4. in a plastic bag, sealed when all the air has been removed - this is the modern way.

Most conservation techniques involve a combined effect on all four factors, (A_w), (T), pH, (O). A good example is fermenting fish (in our Norwegian climatic conditions). A_w is dealt with by the use of salt; T by keeping the temperature below $10°C$; pH by adding some sugar; and O by pressing the fish down under the surface of the brine.

By the way, the 'bog people' of Denmark provide a very good example of conservation in a bog. Here both the low pH (the acidity) of the soil, the low temperature and the lack of oxygen are the conserving factors, so strong that both body and clothes have been preserved for several thousand years.

When all this theory is understood, it is easier for us to put forward the right questions in cases of doubt. If we are given a description of *how* something is conserved, and the description suggests that the conserved product would be unsafe, the explanation may simply be that the informant had forgotten something; so new questions must be asked.

It is also fascinating to use our knowledge of theory to analyse tales told from the past and determine whether they are credible. I have two examples to offer. One involves Attila; the other a boar from the Baltic.

ATTILA RIDING HIS MEAT TENDER
We have all heard the story of Attila, King of the Huns, who came riding with his men from the east and conquered the whole of Europe east of the Danube about 400 AD. He would put a piece of fresh meat under his saddle and eat it uncooked after some days by just cutting off a bit, still on horseback. Can this be a true story? We will use the knowledge of the four decomposing factors and their relationship to the four conserving factors and try to find out.

The meat was safely resting on the back of the horse under Attila's bottom. Here the excess of liquid in the meat is quickly pressed out by Attila's movements. His bottom bumping up and down accelerates the drainage of the piece of meat. If Attila is riding steadily, the water activity will soon be below 0.92, the limit to growth of most dangerous micro-organisms. Probably a piece of leather or cloth has been placed both over and under the meat. These covers absorb any of the drained liquid which does not fly away with the wind. Only the sides of the meat are exposed to free air, they will dry quickly on the surface if Attila rides on, and so he does, you know, not having to stop for eating. No flies will have time to sit down on the sides and lay their eggs; and, if any do, Attila will wipe them off with his knife.

The weight of Attila permits no pockets of oxygen to be left over or under the meat. Thus most of the unwanted micro-organisms cannot find the smallest humid hollow filled with oxygen where they can develop.

The hydrolyzing enzymes, however, and the lactobacilli, both naturally present in the meat, have ideal living conditions created by the body temperature of the horse and of Attila's bottom. The enzymes will ripen or tenderize the meat by dissolving the long protein molecule chains into easily chewable and easily digestible amino acids. The lactobacilli, preferring carbohydrates, will decompose the membranes which consist largely of the carbohydrate glycogen. When this happens, lactic acid is produced, and this increased the degree of acidity in the meat, i.e. lowers the pH. Thus the third conserving factor is brought into the picture besides partly lack of oxygen and reduced water activity. Together they are strong enough to cope with the temperature, favourable to all microbic growth. Thus the story about Attila may be true. When we read elsewhere in the present volume (see Radeva) that the same method was used by the Proto-Bulgarians for hundreds of years, in both war and peace-time, it seems even more likely to be true.

THE BOAR FROM THE BALTIC

At our conference in Poland a story was told about a boar being killed in the Baltic and transported to the south of Europe as a present for a mighty king several hundred years ago. Someone raised the question whether it was still possible to eat it when it arrived. In my opinion it was possible, but to document this procedure may be difficult after such a long time. We can however try to make it up.

It is essential to know how the boar was killed. I believe he was trapped and, when found alive, stabbed in the heart with a knife which had been sterilized in the fire. He was stabbed in the same way that the Lapps used to kill a reindeer, by cutting the main vein (aorta) in the heart, without immediately killing the animal. (The heart will continue to beat for some time, thus emptying the body completely of blood. Since blood is a substrate for the growth of many micro-organisms, it is essential to get rid of the blood quickly and thoroughly, if meat is to keep well.)

When dead and gutted, he was skinned very carefully; there had to be no cuts that could be a source of infection. (The meat inside the membranes is usually sterile in a healthy animal. If the membrane is cut during skinning, infective bacteria from the knife, the hunter's hands, or the surroundings may penetrate and start a dangerous infection in the meat.) Then the surface of the boar, inside and out, was treated. Salt and herbs were rubbed into the membranes, and some freshly cut juniper filled into the empty belly. (The rubbing of the surface with salt lowers the water activity in the membranes, i.e. makes them dry and thus prevents growth

Kirsten Berrum '88

of several bacteria.) If the boar was in addition rubbed inside and out with the leaves of stinging nettle, the formic acid in the leaves would have lowered the pH and thus introduced another conserving factor.

Something would have been done to protect the surface from bacteria in dirt and dust during transport. I believe they wrapped the boar in clean strips of linen like a mummy, to keep dust and dirt as well as oxygen away from the meat, and also to keep the whole animal in a kind of vacuum. In this way he was well protected from attacks by micro-organisms. However, the beneficial decomposing elements, hydrolytic enzymes and lactobacilli, were not prevented from doing their work, and could do as good a job on the boar as on Attila's meat.

Thus transported on horseback in winter, changing horse and man continuously, the boar arrived in the king's kitchen after two to three weeks. Here he was roasted on a spit, and the king was served a tender and juicy joint of meat, better than anything he had had before.

This is a fairytale, but it may just as well have been reality.

CONCLUSION

The Nordic people have, for several hundred years, practised a method of preserving fish by burying it in the earth. This may sound almost incredible, just like Attila's meat and the Baltic boar. However, since 1348 we can find bits and pieces of description of this tradition in the annals, quite enough to establish the practice as fact. What has been more difficult has been to explain exactly how and why the practice was done. The sources are not sufficiently reliable and informative to permit conclusions on these points.

However, a book was recently published about Icelandic coastal culture. Here the author, Ludvik Kristkjanson, gives a detailed description of this traditional method, still in use, for preservation of Greenland shark. Using the theoretical knowledge given in the present paper, it was possible for me to analyze the procedure step by step and to explain both method and purpose. (This analysis is published in the Oxford Symposium *Proceedings 1984-85* and will not be repeated here.)

The conclusion must be that the more precise and detailed the description of *how* a process of food preparation for conservation is performed, the greater chance there is to find out *why* this and that are done.

REFERENCES

Andersen, Erner, *Industriel Levnedsmiddelconservering,* bd 1, 11 and 111, Kobenhavn, 1965.

Asbjornsen, P. Chr., *Fodemidlenes opbevaring i Land og By,* Christiania, 1861.

Berg, Gosta, 'Rotk skinka, torkade gadder ock surstromming' in *Svenska kulturbilder,* bd 6 (old series), 1932.

Eknaes, Asmund, *Innlandsfiske,* Oslo, 1979.

Erichsen, Alf, 'Konserveringsmedlarnas historie', typescript, SIK, Goteborg, 1979.

Hjorthoy, H.F., *Physisk og Oekonomisk Beskrivelse over Gudbrandsdalen Provstie,* Kobenhavn, 1785.

Hofman, Dr. K., 'Die Einfluss der Konservierungs verfarhen auf die Qualitat von Fleisch und Fleischerzeugnissen', in *Die Fleischwirtshaft,* 11, 1972.

Jarvis, Norman D., *Curing of Fishery Products,* Research Report nr 18, United States Department of the Interior, 1950.

Ludvik, Kristjansson, *Islenzkir sjavarhattir,* Reykjavik, 1983.

Marvik, Steinar, 'Vannaktivitet i kjolekonserver' in *Tidsskr. Kjemi, Bergvesen, Metallurgi,* bd 36, nr 10, 1976.

Marvik, Steinar, *Breakdown of added sugar during ripening of sugar-salted herring* SIK Service-Serie nr 664, Goteborg, 1979.

Johan-Olsen, Olaf, *Om Opbevaring av Levnedsmidler til Husbrug,* Kristiania, 1904.

Pledge, H.T., *Science Since 1500,* Ministry of Education, London, 1939.

Riddervold, Astri and Andreas Ropeid, 'Popular Diet in Norway and Natural Science during the 19th century, in *Ethnologia Scandinavica,* Lund, 1984.

Riddervold, Astri, 'Konserveringsmetoder for kjott, fisk, ville baer og urter', Magistergradsavhandling, Institutt for Etnologi, University of Oslo, 1978.

Skjelkvale, Reidar, 'Vannaktivitet og vannaktivitetsmalinger i naeringsmidler', Norwegian Institute for Food Research (NINF), undated typescript.

Smith, Axel Christian, *Beskrivelse over Tryslid Praestegjeld,* 1784.

Tjaberg, Tore B., Vannaktivitetens betydning ved produksjon av spekepolse, Norwegian Institute for Food Research (NINF), informasjon nr 3, 1974.

THE WAYS OF FOOD PRESERVATION IN SLOVAKIA

Rastislava Stoličná

This article deals with the basic techniques used in Slovakia for preserving foods, and with trends of development and changes observed in them from the end of the 19th century up to now. In the rural environment foods of animal as well as vegetable origin were preserved; among the former, meat and milk foods, and among the latter, fruits and some kinds of vegetables.

MEAT PRESERVATION

The kind of meat most frequently eaten in Slovakia till the end of the 19th century was mutton. In the southern regions, however, pork was more often consumed, although in other parts of the country it was eaten only sporadically by the more prosperous people. Beef was eaten only here and there in villages, when an animal was injured and had to be killed or if there was a possibility, it was bought from butchers in the city.

Poultry meat was also rare in the diet, appearing mainly as an attribute of ritual food. Fish was eaten quite often, especially in riverside localities.

Fresh meat was rarely prepared in the villages. Mostly, it was only boiled in salted water, or roasted on the spit or in the embers of the fire. Most meat was consumed in preserved form, so that a supply was available for a longer period.

Among the oldest ways of meat preservation are salting and smoking. Salting was usually a component part of several preservation techniques. Most frequently it preceded smoking. Originally, meat was salted in a series of layers and stored in a wooden vessel. This technique was used especially for mutton and pork, but was applied to beef, goat meat and venison as well. Salting poultry meat to prevent decay has been practised in the villages up to the present time. Salt fish bought in markets or shops was also popular.

By the beginning of the 20th century, under the influence of butchers' practice, meat was being preserved by pickling, laid in a solution of salt, water and seasoning. This technique was mainly used as a preliminary to the smoking of pork.

In the past, the space under the roof near the chimney served for smoking in the country, but during the last few decades, special, independently located, smoking chambers have been built. Smoking was used to preserve mutton, goat meat, pork and also fish. Even in the 20th century preservation by smoking is generally practised, especially in processing pork and pork products, e.g. sausages, ham and bacon.

In Slovakia there used to be other, well-known and archaic, techniques of meat preservation. One of them was drying meat in the open air. In this way mutton and goat meat were dried, elsewhere also pork and fish. But this technique has ceased to be practised in the 20th century.

For short-term storage in the past, meat was dipped in running water or in a well. To strengthen the preservation effect, it was salted and packed in vegetable leaves, e.g. of nettle or burdock. Meat thus prepared was sometimes stored in a cool room or a barn. This technique gradually died out in the 20th century. We have evidence only from the way of life of shepherds, fishermen, etc.

Another old way of protecting fresh meat was to use pits insulated by straw and lined with ice, in which meat was put in winter. To pack fresh or smoked meat in wooden vessels and bury these in soil or in ashes is another archaic method. In the cold winter months meat was also frozen, but in the 20th century this method has only been used for the short-term preservation of venison. (When stored for a longer period, venison is still packed in vinegar and oil.)

Preserving meat in the fat obtained by roasting it is a younger preservation technique in Slovakia. It is mainly used for preserving pork.

At the present time, one must say that the traditional ways of preserving beef, mutton, goat meat and fish have all fallen into disuse. However, the techniques for preserving pork and pork products survive; and there are also new ways, for example, sterilization in glass jars or tins. In addition, the freezing of meat in electric freezers is a phenomenon now generally found in the Slovak countryside.

PRESERVING MILK PRODUCTS

The alpine and submontane regions in Slovakia were characterized in the past by cattle and sheep breeding, so milk was one of the fundamental elements of folk diet. Butter and curd were made from cow's milk in the households. The curd was more often eaten fresh, but sometimes processed for preservation by sun-drying or smoking in a roofed space.

In northern Spiš (North Slovakia), dried curd was pressed into a pot filled with wine or some other spirit. We also have evidence from here that curd was mixed with mashed cherry-leaves and yeast and left until half rotten.

In Turec (Central Slovakia) curd was dried until it hardened completely: this small triangle-shaped cheese was called *tur*.

'Cheese' curd was also made of sheep's milk. By kneading this curd (with wine and mashed wild thyme), *bryndza* was made. Then it was pressed in wooden vessels until ripe and ready for distribution. This aromatic product was known as a speciality outside Slovakia as well. Saffron, caraway, nutmeg and fresh butter were also put in *bryndza*. For long-term storage, it was usually covered with a layer of butter.

Apart from *bryndza*, so-called *ostiepky* were made from sheep's curd in Slovakia. Kneaded pieces of curd were dipped in warm water or in sour milk and pressed into wooden carved forms, the shape of a big egg. The *ostiepky* were then taken out and dipped in salted water to harden and be preserved. Finally, they were smoked. Smoking was also used to preserve other products of sheep's curd called *parenice*.

PRESERVATION OF VEGETABLES

Among vegetables, cabbage was preserved most frequently; and this, together with legumes and potatoes, formed the basis of vegetable diet in Slovakia. In Slovakia, it was grown mainly in the less productive alpine and submontane areas.

Most of the cabbage was, and still is, consumed in rural districts, where it is chopped and pressed into barrels for preservation. Fermented sour cabbage can be stored for a longer period in sealed vessels and enriches the diet all year round.

When appropriately stored, cabbage-heads can be preserved as well. They are kept in cold cellars in and outside the house, in compartments, or they are earthed.

In Slovak folk cuisine, sour cabbage was used in the preparation of various dishes, e.g. soups and thick gravy, and could be mixed with boiled noodles. Stewed cabbage was used as filling for pastry and was also added to noodles, etc. Cabbage is still widely used in these ways in Slovak villages at the present time.

In the first decades of the 20th century, 'black mustard', *Brassica nigra*, was preserved here and there in Slovakia because of its high resistance against cold. The leaves were dried for several days until they had started to wither, then separated, put on strings and sun-dried. On these dried leaves, bread and flat cakes were baked. Black mustard, like white cabbage, used to be cut and preserved in barrels. However, it is no longer grown in Slovakia.

At the turn of the century, horse-radish was commonly conserved. The leaves were chopped and fermented like cabbage. The now sour leaves were boiled and thickened and eaten instead of cabbage.

PRESERVATION OF FRUIT

Among the oldest techniques are jam boiling and fruit drying. Preserving fruit by boiling was always of great significance in rural districts. Jam was made especially from plums for domestic consumption, and was also distributed from fruit-growing areas. In submontane regions in the past, jam often made up for bread and fat as a daily component of the diet. Women had to master its preparation as perfectly as bread-baking.

Before being boiled, plums were stoned and squeezed through a sieve. (The dried stones were crushed, and the kernels were pressed to make oil that was mainly used in fasting periods.) Jam was boiled in the kitchen on an open fireplace or in small temporary stoves built in the vicinity of the house. A fireplace was dug in the ground, bricked on one side and supplied with a copper. The smoke from the stove was driven through a tin chimney. In some places, a tinned recipient was used instead of a copper, so that a greater quantity of jam could be boiled at the same time. The mass was stirred with a wooden spoon or with more complicated mixing devices. The jam was ready when it started to separate from the spoon. It was poured, still hot, into earthen, wooden or straw vessels and, after turning solid, it was turned out. The hard jam was packed in paper and hung under the roof, like bacon or ham. The whole pieces of jam sold well, particularly in the Christmas season. Now, jam is still boiled in households, but mostly in metal barrels with chimneys and only for family consumption, not for sale.

Drying of fruit, especially of plums, pears and apples, also has a long tradition in Slovakia. Evidence of this is provided by its popularity in Vienna markets as well as the big reserves of dried fruits found in the estates of servants of the 18th and 19th centuries. The decrease of dried fruit production from Slovakia was caused by competition of plums from Bosnia at the beginning of the 20th century.

The picking and drying of the fruit represented the culminating point of the whole year's work. Fruit was treated in special drying houses, especially when the farmers could not sell it fresh. The special drying houses have a long tradition in north-western and eastern Slovakia.

There were various different forms of drying house. The simplest design

OVEN WITH DUCT

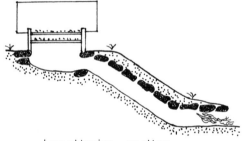

Lengthwise section

Cross section

Ground plan

FRUIT DRYING HOUSE
ORIGINATING IN THE YEAR 1850

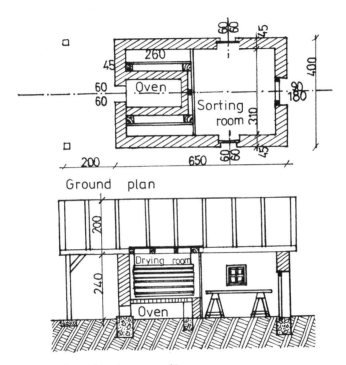

Ground plan

Lengthwise section

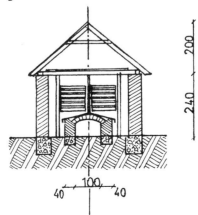

Cross section

was a subterranean stove consisting of two parts: an oven and a fireplace, connected by a duct (figure 1). The oven was a shallow pit, with sides covered by stones, and a wicker frame with the fruit laid on top of it. A subterranean duct led from a fireplace to the oven. When dried, the fruit was covered with a plank coated with mud which was in turn covered with cloth.

The most advanced drying rooms were built in fruit-growing regions. In a log-hut or stone building, with a ground-plan consisting of three parts, there was provision for the drying oven, the frames for drying, and the activities connected with sorting the fruit and loading it onto the frames. (See figure 2.)

Elsewhere in Slovakia, where fruit was less plentiful, drying could be done by the sun, in smoke rooms, or in a bread oven. Fruit was dried by sun in front of the house in protected places, having first been sliced and put on paper. (Smaller amounts were dried on strings.) The archaic way of drying in black smoke rooms (which ceased to exist after modern methods of heating houses were introduced) involved the fruit being put on frames under the ceiling beams so that it was dried in the ascending smoke. Drying in bread ovens usually occurred in placed where the heat from baking bread and pancakes was available for drying. This way of drying was also applied to mushrooms, various kinds of healing plants and seasonings. Fruit and other foods are still dried in this way in villages, often now in metal stoves or electric ranges.

Various dishes used to be prepared from dried fruit, e.g. a milk soup was cooked of plums and cherries. Dried plums were added as a piquant seasoning to bean soup. During fasting periods a sauce was prepared from dried cherries. In winter they often boiled dried fruit to make a 'pap', which in case of need could be replaced by paps of cereals and legumes. Dried plums and pears used to be chopped, mixed with sugar and used as fillings for various cakes. Jam was spread on bread, and also used as a filling for cakes, but it was often used with noodles too. Juice from boiled dried fruit was also in use; in case of need, it replaced milk, it was poured on noodles, and it was sometimes consumed instead of beer. Nowadays, these dishes occur only sporadically in rural regions. Fresh fruit is now eaten to a greater extent.

CONCLUSIONS

It follows from the above that a wide variety of preservation techniques was applied in Slovakia in the past. Home-cured pork and pork products can still be found. Fruit and various plants are dried in households to a smaller extent than it was in the past. Cabbage fermentation is still a general phenomenon.

Apart from the old ways of preservation, new, more complex techniques, e.g. fruit canning, sterilization, pasteurization, were applied after World War II. At the present time, preserved foods (particularly meat, meat products and sauerkraut) represent the major part of the foods produced in Slovak homes.